EDITORIAL REVIEWS

"The next crop of Product Managers can no longer rely on principles of the past. They need a fresh set of guidelines to build a career in this era, to internalise what empowerment means in modern orgs, and the inevitable impact of AI. Bart, Teresa, and Diana have done the product community a big service with Next-Gen Product Management by shedding much-needed clarity on these areas. Highly recommended for someone starting out but also for folks who have been losing their spirit and belief in the discipline. The PM role isn't dead. It's evolving and this book shows why."

— Aatir Abdul Rauf, VP of Marketing at vFairs and Author of the newsletter *Behind Product Lines*

"Next-Gen Product Management provides helpful perspectives on the discipline of product management, ranging from its origins, past emphasis, and current focus. It covers how to prepare for a product management role and get your first job in product as well as how to move up in responsibilities and compensation, including senior leadership. I especially found the "five behaviors to practice as a product manager" valuable for long-term success. Each of the three authors provides their perspectives on what successful product managers do, providing a breadth of knowledge. Also, a product management book would not be complete today without including how AI tools are used by product management, which is also addressed. This is a practical book of proven practices based on the authors' experiences and what they have seen others put into action."

— Chad McAllister, PhD, Product Management Professor, Practitioner, and Host of the *Product Mastery Now Podcast*

"Next-Gen Product Management is an essential resource for anyone building a product management career. The authors discuss the history of the role, how it's changed, and provide a vision for where it's going in the future - including the impact AI will have on how PMs do their work and what products they build. Everyone, from first-time PM to VP of Product, will find valuable insights to accelerate their career."

— Ravi Mehta, co-founder of Outpace, ex-(Tinder, Facebook, TripAdvisor, Microsoft)

"Next-Gen Product Management provides valuable insights into the evolution of product management, offering foundational skills for early-career professionals and fresh perspectives on generational diversity, collaboration, and stakeholder engagement for seasoned leaders. I found its focus on generational dynamics insightful. It's a powerful reminder that building great products requires not only understanding but also fostering diverse, high-performing, and creative teams. As leaders, it's essential to create environments where collaboration thrives, allowing us to embrace unique perspectives and evolve our ways of working to deliver the next generation of products."

— Carmen Palmer, CEO of Women in Product

"The craft of product management has evolved drastically over the past few decades; each new wave of emerging technology, from the ubiquity of internet access to the accessibility of cloud computing to the proliferation of mobile devices, has forced the evolution of operating models for PMs across domains and levels. In Next-Gen Product Management, the authors have done an admirable job of outlining the major milestones in the rich catalog of product development frameworks, not only highlighting what makes certain practices timeless, but also illuminating how the next generation of product leaders can future-proof their careers."

— Ibrahim Bashir, Product Executive ex-(Amazon, Twitter, Box Amplitude) and Author of the newsletter *Run the Business*

"This book hits different. It's like upgrading from a used car to a rocket ship in your PM career. The authors break down old-school product management approaches and show how today's top leaders are actually getting things done. Spoiler: it's not about playing the "mini-CEO" anymore. From shaping roadmaps to using AI tools like ChatGPT, it's packed with practical, modern strategies you can apply right away. What I loved most? Their personal reflections. It feels like sitting down for coffee with someone who's been in the trenches and knows what works. No fluff, no boring business jargon. It's simply real, actionable insights. If you're serious about stepping up from PM to true product leader, this book is the shortcut you need."

— Moe Ali, CEO of Product Faculty

"A must-have reference for Product Managers at all stages from aspiring, to entry-level to rising leader. The role of the Product Manager is broader than ever before. Next-Gen Product Management brings much-needed clarity to the fundamentals, drawing on decades of real-world experience and a deep understanding of the trade."

— Guarav Vohra, Startup Advisor, Superhuman founding team

FUTURE PROOF YOUR CAREER

NEXT-GEN PRODUCT MANAGEMENT

TERESA CAIN   DR. BART JAWORSKI   DIANA STEPNER

Lucid
Creative Press

https://www.lucidcreativepress.com

From Bart: To my wife. This wouldn't happen without you.

From Diana: To my husband, for his unwavering support.

From Teresa: To my kids and husband who inspire me to write each and every day.

Contents

Introduction 1

PART I: PRODUCT MANAGEMENT IS CHANGING,
BETTER KEEP UP .3

Chapter 1: Product Management "Then" vs. Product
Management "Now" 4

    Product Management *"Then"* 7

    Product Management *"Now"* 9

Chapter 2: The Mini-CEO 12

    What Did it Mean to be a mini-CEO?. 12

    Does the mini-CEO still Apply? 14

    What are other Ways to Lead? 17

Chapter 3: Everchanging Product Frameworks 19

    Foundational Influences. 20

    Modern Frameworks 21

    Small vs. Large Organizations 22

Author Reflections: Teresa Cain 24

    My Favorite Product Management Certifications. 25

Chapter 4: Your Path to Product Management. 29

Four Ways to Launch a Career in Product Management. . . . 30

Landing a Junior Position 33

Landing a Promotion 36

Chapter 5: Challenges for Product Managers to Overcome . 38

Challenge #1 Remote, Hybrid or In-Person 39

Challenge #2: Competitive Salary 40

Challenge #3: Tech Layoffs. 40

Chapter 6: What Does it Take to be a Successful
Product Manager Today? 42

What are Poor Behaviors to Stay away from...and Why? 44

What Characteristics Should You Embrace? 48

PART II: FUTURE PROOF YOUR CAREER AS A
NEXT-GEN PRODUCT MANAGER 52

Chapter 7: Where is the Next Evolution of
Product Management? 53

What Can You Do to Prepare? 56

Chapter 8: Practice, Don't Preach Product Management
Principles 60

Managing Your Roadmap 61

Managing Your Backlog 65

Creating User Stories 69

Author Reflections: Dr. Bart Jaworski **72**

How I Broke into Product Management 73

Chapter 9: Be a Product Champion, Not Just an Order Taker **75**

Start with Empowerment 75

How has Empowerment Changed? 77

How Do You Create an Empowered Team? 79

Exercises to Encourage Empowerment 81

Establishing a Community of Practice 85

Chapter 10: Drive More Results with Data **89**

Start with the Right Questions 90

How to Get the Best Data 93

Chapter 11: Using AI Tools to Stay Sharp **97**

Streamline Research with ChatGPT 98

Improve Writing & Communication with Grammarly 98

Become a Better Speaker with Orai 98

Get Predictive Analytics & Deeper Insights with Amplitude. . 99

Create Workflows & Roadmaps with FigJam 99

PART III: PRODUCT MANAGER TO PRODUCT LEADER. **101**

Chapter 12: Success for a Product Manager vs. Product Leader **102**

How to Measure Success? 103

Chapter 13: You Don't Need to Be Loud to Lead 106

Leadership Styles .108

Create an Inviting Space for Your Team 110

Author Reflections: Diana Stepner. 114

How I Approach Leading a Big vs. Small Team 115

**Chapter 14: Using Your Leadership Style to Deliver
Results. 118**

Why Am I Talking? 119

Aligning Different Types of Leadership Styles 121

Establish Trust 122

**Chapter 15: Managing Stakeholders at Every Level of the
Organization 124**

Keeping Sponsors Happy 128

Balancing the Worlds of Sponsors and Users 132

Conclusion 136

What has Changed in the Workplace Culture? 136

What Considerations Need to be Made Today? 139

Top Skills to Stand out in Today's World 141

What Stands out for Product Leaders? 144

What Stands out for Product Managers? 147

How Many Stars Would You Rate This Book? 150

About Teresa Cain. 151

About Dr. Bart Jaworski 154

About Diana Stepner 156

Endnotes. 159

Special Thanks 169

INTRODUCTION

Do you know when product management began? It's likely longer ago than you expected. The concept of product management can be traced back to 1931 during the Great Depression when a then-junior executive named Neil H. McElroy at Procter & Gamble drafted a memo that teed up a role for a new position in the company.[1]

It shifted decision-making closer to the customer. He titled the memo *"Brand Men"* and assigned tasks to manage and promote the product, track sales and solve problems for the customer.

The title of the memo exposes the absence of women working in product management in the first half of the 20th century compared to today. Women taking on leading roles in this industry is only one of the changes in product management over the years.

However, the biggest transformation is not whether men or women lead the product team. The biggest shift is generational, for product leaders and their employees. Product leaders have their own innate generational mindset and also must finesse generational differences among their team members from those in the Silent Generation to Baby Boomers, Generation Xers, Millennials, and Generation Zers.

While we each have our own story to tell about how we entered the product space, the entry to the market as a product manager has evolved significantly over the last couple decades.

Getting hired as a product manager used to be challenging to do fresh out of college without having a few years of work experience. In fact, if you compare job descriptions for a product manager role 20 years ago versus today, having an MBA was once a requirement but now is considered only a preferred qualification.

What's even more remarkable, it's possible to get hired as a product manager right out of college, and you may not even need a bachelor's degree. However, you will now have more competition than ever.

The product management market in our post-pandemic era is saturated. If you're not already in a product role, jumping into one now may be tough without the right guidance. That's why we are writing this book!

Throughout our journeys in product management in the last two decades, we took on new roles across products, leading teams for digital and physical products in B2C and B2B.

Together we offer three generations of product management expertise to help you navigate the world of product management, whether you're just getting started or well into your career.

PART I: PRODUCT MANAGEMENT IS CHANGING, BETTER KEEP UP

CHAPTER 1

PRODUCT MANAGEMENT "THEN" VS. PRODUCT MANAGEMENT "NOW"

Product management of a software product, just like the technology it develops, has evolved significantly in the last 20 years. Starting out as dreamers, inventors, and pioneers, most of us are highly specialized craftspeople pushing the envelope to achieve potential improvements in metrics.

In this chapter, we compare the challenges and landscape that confronted the *"founders"* of product management with what one can expect from the position today.

Is it better to explore the uncharted frontier of the digital landscape, having little idea what works, what doesn't, and how to get there? Or, is it more difficult to improve upon what's become second nature and par for the course in the information technology *(IT)* world?

Let us start by saying we believe the product management space experienced a journey similar to what every successful company endures.

You begin small, with an idea of how to solve an actual user challenge.

You have limited time and budget and a bunch of passionate teammates ready to make it happen.

However, you also need revenue, so you need to release the product as soon as possible. Thus, you cut corners. At this point, you don't really care about scalability, user privacy, or code quality.

In this stage, a CEO might say, *"Let's have it at least working and we'll fix the issues as they come."*

Thus, the product launches, it lands some initial clients, bugs are found and fixed, and all appears good. The team and product start to grow.

Then, the potential risks and issues slowly start to pop up and are dealt with amidst various levels of pain, public backlash, and product rewrites. After years of devising internal processes to battle any potential setbacks, you end up with a gigantic company that's extremely difficult to navigate.

Instead of being nimble and flexible, innovation encounters lethargy and clunky processes. Developing anything takes ages.

Sound familiar?

Truth be told, product management and the entire IT industry started as a niche, boutique business. Various sins, shortcuts, and omissions would be accepted because they would affect a limited number of passionate users.

For example, consider the issue of email privacy. Just a few years ago, signing up for any product or service with an email address meant you involuntarily opened your inbox to a few marketing emails.

At worst, you received a flood of spam. Your email address could be freely traded between companies and then bam! The situation became so bad the government stepped in and the *"privacy revolution"* began.

Nowadays, companies can't send promotional emails without explicit consent and the data holders are mandated to keep it safe and private.

While this protects user rights, it also makes marketing and day-to-day operations for product managers much more difficult. Regulatory compliance is one of the reasons your teams could spend weeks and months in preparation before releasing new product updates.

Our little niche slice of the IT world is now mainstream. People's lives and livelihoods are at stake, even for the smallest of products.

As popularized by Spider-Man's uncle in the Marvel comic book series *(but as Voltaire famously said)* **"With great power comes great responsibility."**

You feel the weight on your shoulders on nearly a daily basis whether you work at the biggest IT companies in the world, like Microsoft, or the smallest startup looking to make it big.

A product manager's role 20 years ago was quite different than it is today. How is that, you ask?

Product Management "Then"

When we think of being an early product manager years ago, when digital products became more popular and widely accessible, two things come to mind: risk and the unknown.

No product manager knew what would work as a concept as there were no real discovery patterns on which to fall back.

No one truly knew whether a specific problem was worth solving or if it would be profitable. Apple provides a great example to illustrate.

Initially, the Personal Computer *(PC)* concept was presented by Steve Wozniak to his managers at Hewlett-Packard *(HP)*. Result? The idea was rejected multiple times; no one at HP bought into the concept that every day Joe would purchase a computer.

This ongoing stubbornness ultimately led Wozniak to realize his vision with Steve Jobs. The rest is history. Mr. Jobs really lived ahead of his time and understood what the masses would like. Just like when he was inspired by the mouse-operated, icon-based operating system interface from a research facility at Xerox.

But we digress. What we mean to point out is that early product managers had a clear runway to be true visionaries. The sky was the limit to amassing fortunes and building society-defining IT products from literally nothing.

The market basket of opportunities was literally undefined! However, just like being an explorer venturing into the wild, there was a major disadvantage: the road of unknowns, laying ahead in the darkness.

Developing quality software is and always was challenging. Early product managers had much less freedom to make mistakes and experiment. Only some people had access to the internet, and physical distribution of updates was still nascent.

That meant your first release was likely your only release. In such conditions, the waterfall development model *(where you start with a detailed plan of the end product and follow it to the letter)* was the only viable option and that meant you only had one shot to do it right.

Of course, there were some versions of the feedback loop. However, any discoveries based on scarce user feedback could realistically only be implemented in the next major version of a specific product. Remember when an update of Microsoft Office or Photoshop would pop up almost every year?

We opened this section by saying early product managers were dreamers. But that wasn't limited to brainstorming a product idea. Forget the client feedback loop! There simply was no data available to inform decisions.

Back then, we lacked access to live product analytics tools available today, such as Amplitude. You didn't know how your product was used, what were the most popular features, and what bugs were happening.

At least that wasn't available until the cloud software distribution model became more popular. Thus, product managers had only one real metric to follow: sales. If your software delivered that, that meant you would get to work on its next version or at least have management confidence in your next project.

Of course, there also were issues in using optimal processes, as those were not yet discovered.

There were times when the developers would need to manually work out how to update the code they were working on together, as there was no such thing as version control software like Git.

Though this isn't strictly product management, we wanted to make a point that code development took longer in the past. However, when comparing development times early in our career to current ones, we admit they didn't change too much.

What changed are the steps to get from an idea to completed, live code. That said, it still feels like it takes way too long. Since we mentioned how code development has evolved, let's use this as a segue to the next section.

Product Management "Now"

If managing a product *"then"* was like venturing into an unknown frontier, we would compare today's landscape to taking the same bus to work every day. Of course, this is putting it to the extreme, but hear us out.

You'd think by now we would be operating in paradise for product management. The internet and its products reach basically every corner of the planet and even beyond. The whole world has become your market and the opportunities are boundless like the mantra *"to infinity!"*

However, reality keeps us tethered to the real world. As the product manager, you can see everything users are doing, where they click, and how long they use the product.

You can iterate and get the newest version of your product to clients and users whenever you wish.

You have access to thousands, if not millions of pieces of feedback and reviews, not to mention companies dedicated to helping you with product discovery.

Whatever you can come up with as a product idea, you can research in multiple ways and have confidence based on data that you are making the right choice before a single line of code is written.

But…

It has become extremely difficult to innovate. Arguably, data overload led to innovation overboard. The rise of large language models *(LLMs)* and AI technologies in late 2022 shook things up and opened up the world of possibilities for many companies and products. But for us, reality paints a grim picture.

The mantra *"to infinity"* has become *"been there, done that."* For one thing, years of unlimited range allowed innovative ideas to find their market niche.

If we exclude the new AI focus in product development, it seemed as though one could either create new products by introducing an old solution in a new form *(say a cheaper, lighter, luxurious, or local version)* or combine similar products into one *(i.e., a tool that offers a bank services and fees comparison engine)*.

In recent years, we saw a few *"hype"* technologies enter this space: web3, crypto, NFT, AR/VR, and finally LLM AIs. They represent the next generation of innovation, promising new frontiers in information technology, giving product leaders hope that new discoveries could rise to the levels of Google or Meta.

An alternative is to look for older technologies and give them new life using modern hardware. An example here would be Virtual Reality headsets that already existed in the '90s! Trust me, playing Doom II on one of those headsets cost $1 per minute and left an impression for years to come.

Albeit, they are a piece of junk compared to modern headsets from Meta and Apple.

There is another negative angle to the current, hard-to-innovate landscape. In private conversations, many product managers admit they change the product for change's sake, to justify upping the cost to customers every year.

However, most of their work doesn't make a lot of sense or represent a terrible return on investment *(ROI)*, by barely, if at all, moving the metrics needle.

So if there is little space for real innovation, how should product managers invest their time and effort? Today, it's all about looking into existing solutions at competitors and other industries. Sometimes this works nicely.

For example, the stories feature copied from Snapchat is a

resounding success on Instagram, but it was a complete bust when introduced to Skype. Of course, it's also about polishing the products as well. Make them look more modern, quicker, slicker, and user-friendly.

With tons of data, the world of unlimited possibilities moved from innovation to ongoing improvements.

This leads us to a more positive aspect of modern product management. Many tools help day-to-day, optimize processes and have been formalized years ago. Technical shortcomings have been mostly overcome.

This means that product managers can focus on research, creativity, and user perspective rather than figure out how to jump through technical hoops. If a product manager figures out what can improve the metrics and user experience, it's most likely achievable.

CHAPTER 2

THE MINI-CEO

In the early days of product management, an effective product manager commonly was referred to as the mini-CEO of the product. Product managers were a rare breed and training programs were just starting to form.

At the time, the designation made sense. Just like a CEO, single product managers often had authority over all aspects of their product or product line and directed multiple functions to deliver on those goals.

What Did it Mean to be a mini-CEO?

Just as CEOs execute their jobs differently, one would expect variations on the mini-CEO as well. At technical companies whose teams provided behind-the-scenes functionalities, you would find more technical product managers who honed their craft from the engineering ranks.

In organizations whose products were consumer-facing, a business savvy product manager likely called upon expertise in marketing and strategy to bring products to market.

At enterprise organizations, the product manager often rose through the ranks from the sales channel to lean in on customer insight and relationships. In the pioneering days of product management as a discipline, it was common to see product managers remain at one organization – *and consequently in one industry* – for an extensive period of time.

They were building and honing their expertise in the company as they advanced through the organization, in the same manner as a CEO.

Google and other organizations introduced rotational programs to train early career product managers. Consequently, these programs knowingly or unknowingly developed actual CEOs.

Product managers such as Kevin Systrom, Marissa Mayer, Satya Nadella, and Sundar Pichai[1] moved into leadership roles. Their cross-functional knowledge and experience in business, as well as technical acumen, served them well as they climbed the corporate ladder.

These individuals illustrate a professional template for self-leadership, focus on continual learning, and adaptability that it takes to effectively lead a product and an organization.

Even though these characteristics remain critical components for an effective product manager, the mini-CEO designation has fallen out of favor.

Today, more collaborative and empowered practices have proven more effective for the cross-disciplinary nature of product management.

Just as we have moved away from the stereotype of a product manager emerging from certain schools and executing a specific leadership style, the designation of product manager as a mini-CEO has evolved.

Today's product manager brings teams together and aligns a diverse group of individuals from all backgrounds who contribute, create and collaborate through the different stages of the product development life cycle.

Does the mini-CEO still Apply?

Similar to a CEO, product managers work alongside a leadership team, which consists of their peers in design and engineering. The product manager analyzes the development and potential impact of the product and tracks its trajectory just as a CEO monitors a business.

For example, strategic analysis is done to inform competitive positioning and feature prioritization. Reviews are conducted on a regular basis to ensure desired outcomes are being achieved in a timely manner. Obstacles that could negatively impact progress are addressed to ensure the company goals remain top of mind.

Although product managers apply the collective insight from their cross-functional partners to drive decisions and forge the path forward, they do not have direct authority over the other roles that make up their team.

Whereas a CEO can hire and fire at will, product managers share feedback or negotiate for changes. The final decision is not theirs to make.

Such nuances are particularly important as new generations enter the workforce, and the desire for team empowerment and individual autonomy steadily rises.

Previously, organizations defined acceptable behaviors and hierarchy; today employee input influences processes in the company just as much, if not more – until the actual CEO steps in.

Structuring the role of product manager as a mini-CEO undermines the collective value of design, engineering and other disciplines.

Creating the right conditions for people to do their best work means providing opportunities for different functional roles to take center stage and motivate dispersed and diverse teams.

For example, building on the collaborative power of cross-disciplinary product teams, other functions take the lead at different times in the product life cycle.

For simplicity, for a general product life cycle model, consider the four stage approach: *Discover - Define - Develop - Deliver.* Even though each business will introduce its own terminology, these four stages generally are reflected in some capacity.

Four Stage Approach:
Discover - Define - Develop - Deliver

1. Discover

Starting with Discover, the company strategy informs the product strategy, and consequently prioritizes opportunities for the team to explore.

The company's CEO or CPO, alongside other leaders, will define the company strategy. Typically, user experience research *(UXR)* leads the approach and direction is implemented by the product manager.

At a small organization, a member of the product or design team may fill the UXR role.

If UXR does not exist and there is limited design, the product manager may guide the team through discovery. In all cases, the product manager has travel companions providing design and engineering input.

If the company does not have a defined or consistent strategy, the product manager and cross-functional team members will formulate the product strategy as the company strategy.

(continued)

(continued)

In either case, as Richard Rumelt recommends in Good Strategy/Bad Strategy,[2] the strategy should be defined through a series of questions.

For example, leaning into relevance, how to make it simple and easy for a customer to choose a ***<insert your business activity>*** product?

Or, when a customer has problems during ***<insert your business activity>***, how can you make it easy to obtain assistance?

Framing the strategy as questions enables the company culture to be woven in and makes the strategy more memorable and motivating.

2. Define

At the next stage, Define, the product manager works with the design and engineering leads to narrow in on the problem to be solved.

Team members add value by bringing expertise from their core functions to the table. Engineering is focused on feasibility.

Design covers usability. Product leans into viability and partners with UXR or design on desirability. Ethical considerations should be factored in by all teams.

3. Develop

During the Develop stage, engineering drives the effort. The collaboration with product and design continues with consultation happening throughout the process.

Priorities and future considerations will be driven by the product manager. Design addresses the overarching product experience and addresses questions that arise.

4. Deliver

In a *"drink your own champagne"* analogy, all members of the team are involved in the Deliver stage. Engineering will address bugs prioritized by the product manager.

Design will capture user feedback and improve upon the product experience. Product management, alongside customers, explores all aspects of the product's capabilities.

Whereas the mini-CEO would determine the role and scope of each team member, today's product teams build on the expertise of each member with more fluid ownership.

Different individuals take the lead at various stages of the product development process to bring the optimal product to life.

Today's product teams focused on achieving greatness calls for a team effort. Power is dispersed across team members to enable everyone to excel.

People don't need another CEO, effective product managers motivate people to work together and watch the magic happen when individuals work collaboratively as one team.

What are other Ways to Lead?

Instead of commmanding from above as the mini-CEO title implies, positive leadership character is today's super power for an effective product manager.

Building on the generational influences mentioned earlier, the performance of team members led by individuals who reflect integrity, responsibility, drive, and compassion follows accordingly.

Note that showing compassion and empathy does not mean the product manager is a pushover.

Author Simon Sinek[3] explains when leaders focus less on being *"in charge"* and more on taking care of those who are in their charge, that's a sure sign of a compassionate leader.

CHAPTER 3

EVERCHANGING PRODUCT FRAMEWORKS

Every organization, big or small, adopts its own product management best practices. Likewise, each is modified typically from product frameworks developed over the years.

What's interesting about product frameworks is – *like technology* – they continue to get better over time and have continued to evolve, much like the transition from waterfall to agile, with more flexibility in the process. Consider what many companies do today: *"wagile."*

That means they are somewhere in the middle. Having knowledge of these frameworks is the foundation for becoming a successful next-gen product manager.

We'll dive into four frameworks that have had foundational

influence on technology companies and product roles:

- Pragmatic Institute's courses on how to build and market products *(1993)*;[1]
- Dan Olsen's *Lean Product Playbook (2015)*;[2]
- Marty Cagan's *Inspired: How to Create Tech Products Customers Love (2017)*;[3] and
- Teresa Torres' *Continuous Discovery Habits: Discover Products That Create Customer Value (2021)*.[4]

Foundational Influences

If you look back to the early '90s, the Pragmatic Institute marketing courses really built the foundation of frameworks for what product management is and isn't.

The Pragmatic Institute's framework for building and marketing products has changed over the years but still focuses on aligning product management and marketing activities with market needs and business goals.

This structured approach was built into course and certification options to learn and achieve mastery at different levels including the following:

Foundations: Gain a thorough understanding of your market and the opportunities that drive results.

Focus: Identify and present the right product strategies and get your roadmaps embraced.

Design: Partner with design resources to create innovative solutions that wow the market.

Build: Learn how to align product and development to deliver remarkable products.

Market: Build buyer expertise and create strategic product marketing plans that resonate with the market.

Launch: Align your entire organization around the right product launch strategies.

Price: Learn how to set the right price for each product.

Insight: Uncover new opportunities and trends in your markets and products with data.

Modern Frameworks

The Pragmatic Institute offers training for product managers to connect the products they're building with a go-to-market strategy, something every product manager should be aware of and know how to apply.

Its approach to product management provides the foundation for product managers to understand customer needs and market dynamics, which complements and has influenced other frameworks like those of Olsen, Cagan and Torres.

Respectively, they expand these concepts and put flesh on the bone as they are practiced in organizations they coach. Modern methodologies and practices, as well as access to learning websites like Udemy.com[5] have arisen and made access to product management founding principles more accessible.

Whereas the Pragmatic Institute laid the foundation for Olsen, Cagan and Torres, each of their frameworks is different and has a tailored approach on how to be successful when building products. Here's how these three frameworks are unique:

Three Product Frameworks

1. *Lean Product Playbook* by Dan Olsen

Olsen's Lean Product Process emphasizes a structured approach to product development, focusing on understanding customer needs, validating ideas with

(continued)

(continued)

Minimum Viable Products *(MVPs)*, and using data to drive product decisions. His methods are designed to help product managers create products that are both innovative and aligned with market demands.

2. *Inspired* by Marty Cagan

Cagan's approach to creating products customers love emphasizes the importance of understanding customer needs, empowering teams, and creating a clear product vision and strategy to drive innovation and build successful products.

3. *Continous Discovery Habits* by Teresa Torres

Torres' framework provides a practical guide for product teams looking to improve their ability to continuously discover and deliver products that solve real customer problems. She outlines a framework for embedding customer feedback into the product development process through continuous discovery, which involves regularly conducting customer interviews, running experiments, and testing hypotheses.

Small vs. Large Organizations

The amount of resources available to a product manager *(courses, books, certifications)* now versus two decades ago is like night and day. Product managers today often find themselves creating best practices that cater to how their company wants them to deliver results, versus what may be best.

We have each worked at companies that make up their own

rules for what best practices look like, often with those coming from the experiences of current or past company leadership. Of course, this approach presents a challenge in that frameworks can create structure but also eliminate room for creativity and advancement when you force a framework with boundaries.

For instance, Cagan's book *Empowered*,[6] was published in 2020 and its target audience was product leaders. He writes product managers should spend four hours per day on innovation.

This recommendation itself is a foundational difference between small versus big fish mentality. While there are one million product managers worldwide, a majority are not working for one of the five largest American tech companies.

Top product managers at a MAMAA *(Meta, Alphabet, Microsoft, Apple, Amazon)*[7] may have a $1 million compensation package with base, bonus and stocks. However, a typical product manager at one of these titans of tech might only have a search box as a product and no real way of influencing the organization and its company goals. That's not to say you won't succeed or make an impact.

From our collective experience, the scope of a product manager role at a MAMAA in Silicon Valley comes with the title and pay, but it lacks the depth and ownership you get working as a product manager in a smaller technology company. So what's the difference, you ask? The scope. The scope you get as a product manager in an organization with less than 1,000 or 500 employees is going to be dramatically different.

At a smaller tech organization you have control of the entire product end-to-end, you don't just own a search box, you own every feature within the product and have greater influence due to the close-knit team of stakeholders you work with every day.

This is the same challenge with frameworks. What works well at one organization doesn't provide benefit for another, and most organizations end up modifying what works best for them.

AUTHOR REFLECTIONS: TERESA CAIN

I've had the opportunity to spend almost my entire career in product management, approaching twenty years in this space. During a recent podcast interview, I learned that makes me a *"vicennial"* and not an elder millennial.

I was lucky to have gotten my start in product management after an internship with my future employer. I recall the recruiter told me I was a final candidate for the position. It was between me and a candidate with an MBA.

However, the singular advantage I had over the candidate with the advanced degree is work experience. I was wrapping up my sixth internship, this next one would be my seventh during my college career. I was determined and prepared to land a full-time role upon graduating from college in 2007. Some might say I was fortunate to be in the job market before the subprime mortgage collapse.

Like many in the financial services sector, the Great Recession directly impacted my future employer: H&R Block. Now that I've ended the suspense of my first job hunt, it also foreshadowed the next rung on my career ladder.

The irony of being up against an MBA candidate is I decided then and there that if I did not get that internship, *(which led to a full-time role)* I planned to get my MBA at Northwestern University. I had my eyes on that graduate program in Chicago for years. However, I was eager to begin working after completing two undergraduate degrees in English and Journalism.

My internship with H&R Block shaped my entire career. As an intern on the product team, my first task was to test out competitor tax software and understand the *"it"* quality compared to H&R Block. At this time, H&R Block was losing customers to a big Silicon Valley competitor. Take a wild guess, the competitor was TurboTax. While H&R Block had quality tax software as a self-service option, its wheelhouse was its decades-long leadership in tax expertise and tax preparation services, not for its software expertise.

TurboTax got its start as a software company, not tax preparation. Both companies upped their ante to compete and still give each other a run for their money to this day. By no means is this an unpaid advertisement, so I'll let you choose who does your taxes this year. Needless to say, self-service SaaS software became my passion for every role since then. It has driven me to understand the needs of the customer, and led me to my favorite certifications.

My Favorite Product Management Certifications

If you read my first book *Solving Problems in 2 Hours*, you may recall I worked for an enterprise content management *(ECM)* technology firm from 2008-2012. Perceptive Software was one of the top providers in the ECM space. In 2010, Perceptive Software was acquired by Lexmark International for $280 million[1] and the Perceptive portfolio assets, Perceptive Content *(formerly ImageNow)* were later acquired by Hyland in 2017.[2]

I discussed how the UX maturity at Perceptive Software was ahead of its time. It ranked at Stage 6 of the Nielsen Norman Group UX Maturity Model,[2] defined as a company with dedication to UX at all levels for deep insights and exceptional user-centered design outcomes.

This also tells you where it stood for maturity in the product management space, as well. Perceptive Software was well ahead of the game in product management frameworks when we migrated from waterfall to agile.

It wasn't until my next role that I came upon my favorite framework that you've already learned a bit about, for which I was able to get certified firsthand. While at Perceptive Software in 2012, I was recruited for a project management role at a healthcare technology company, Netsmart Technologies. Netsmart, founded in 1968, is the leading provider of software and technology solutions for community-based healthcare.

If you haven't heard of this company, it would behoove you to learn more about the organization. Netsmart is a billion dollar software company that continues to expand its partnership and client base since I left in 2015.

While here, I didn't stay in the project management role for long. I was quickly promoted to mobile product manager and sent to Minneapolis, Minnesota to complete Pragmatic Institute's Pragmatic Marketing Certification © *(PMC)*.

At the time, PMC meant I completed its series for product management. Over the years, Pragmatic Institute has added many courses and certifications that are mentioned in Chapter 3: Foundations, Focus, Design, Build, Market, Launch, Price, and Insight. Back in the day, training only covered Foundations, Build, Market, Launch and Price.

That in itself shows you how much product management has evolved. In turn, companies like the Pragmatic Institute are evolving alongside the industry to meet the moment. If you are early in your career and thinking about getting certified, this program offered one of the best foundational certifications I've ever completed. And I say this looking through a lens of experience in achieving all kinds of project and product certifications during my *"vicennial"* tenure: Certified Scrum Product Owner, Certified Scrum Master, Project Management Professional and Lean Six Sigma Green Belt.

Alright, so now you know my favorite product management certification for early career, how about mid-to-late career? I consider myself a lifelong learner, so a few years ago I took Northwestern University's Kellogg School of Management Executive Education Certification for Product Strategy methods, a program with emphasis on discovering, developing, managing and marketing products as a business. I highly recommend this four-month course as a refresher to breathe new life into your perspective.

I will tell you, it was invigorating to see what's changed as I've progressed in my career as a leader in this space. It also opened new connections with product leaders around the world who have similar roles as mine.

Okay, so now you're wondering, what about late-late career? If you never had the chance to obtain a product management certification, that's okay. If you're looking for a formal option, I've provided two mentioned above.

But today, product managers also can take advantage of online training, such as Udemy. I have courses out there, as does Dr. Bart, and Diana offers training on Maven.

There are many options for online certifications from top

product leaders. Do your research and take your pick. Alternatively, you could skip all that and read the books we recommend in Chapter 3, from Teresa Torres, Dan Olsen and Marty Cagan. Many organizations build on and pull from their foundational product management expertise. Leaning in and learning from their guidance will allow you to apply it to your job as a product manager.

CHAPTER 4

YOUR PATH TO PRODUCT MANAGEMENT

D o you know what the difference is between a successful and a failed product management aspirant? The successful ones never surrender and ultimately get their dream job. Becoming a product manager is a difficult and weird journey.

This job, even for junior positions, starts with a lot of responsibility, expectations of initiative, and the ability to make data-driven decisions. These are only a few specific skill sets required of a typical product manager. Ironically, all of them are usually acquired during the initial years of one's career.

For example, if a product management junior is hired, this individual typically will work under strict supervision on relatively low-risk tasks. Keep in mind a product manager serves a leadership role shouldering significant responsibilities. Poor decisions can cause the company to lose money, user confidence, and much more.

Four Ways to Launch a Career in Product Management

The product manager position falls into the *"you need experience at the job to land it in the first place"* paradox. Don't fret.

New product manager hires enter the field every year. Let's start with some basics on how to grow the right skills and instincts, often referred to as *"product sense."*

Here are four action items to build out as your launching pad for a career in product management.

Four Ways to Launch Your Product Career

1. Ramp Up Your Product Education

A diploma or certificate will not be the deciding factor in the eyes of a recruiter looking to fill a product manager opening.

That wasn't always the case. However, taking product management classes, whether at a university or through an online institution, is an investment you want to make in yourself and your career.

Such courses will help with a logical, continuous narration and explain the most important aspects of the job. Courses also can be a great starting point to study the position. They will teach you what topics to focus on and what makes a difference.

As authors of such courses, it would be a conflict of

interest for us to recommend or critique any of them. However, we can assure you that you don't need to invest thousands of dollars to learn the right stuff.

There are cheaper and equally effective alternatives. However, no single product management course is enough! You also need to build a foundation, as outlined in the previous chapter.

2. Create a Daily Routine

The best product managers quite simply enjoy their work. Make it a habit to learn the ropes. Be curious long before you land the position. It doesn't need to take up too much of your time.

Simply, reserve at least 30 minutes a day to watch a product management YouTube video, read a book on the topic, or start thinking like a product manager. For that, it will be really useful to analyze the work and projects of others.

3. Read Case Studies and Product Stories

Start by studying the history of core products like Facebook, Google, or Stripe to understand their journey and what made them successful.

The more stories you know, the better you can use them in your future job interviews and line of work. Making references to actions that worked for other products will make you sound amazing during the interview process. For example, when asked about growing a product, casually mention how Dropbox

(continued)

(continued)

grew like crazy due to its gamified referral system. Practice by taking any product you like and suss out answers to the following questions:

- What is its vision?
- Who are its users?
- What are its metrics?
- Who are its competitors?
- What are the key features?
- What are its user personas?
- How can this product grow?
- What is the size of its market?
- What is its monetization model?
- What are the biggest pains to solve?
- What problem is this product solving?
- Are there any untapped opportunities?
- What is the users' rating of the product?
- Are there additional monetization options?
- Why do you like or perhaps dislike this product?
- What would happen if this product shut down today?
- How did it change over the years it's been on the market?
- What would have to change to make this product obsolete?
- How did the users address the same problem before having that product?
- What would you look into first if you were the product's newest product manager?

You can use ChatGPT, i.e., in the form of a Copilot bot in Edge, to share your answers and see if AI agrees with you.

It might not be ideal, but if you don't have access to a product manager to check your answers, this is the next best thing.

Speaking of the best thing to do for aspiring product managers, be sure to eventually dedicate time to build your own product.

4. Build Your Own Product

That's right! This is a great way to use your skills and make something successful to showcase. Of course, you don't need to create an entire business, something small will do, such as a blog, a YouTube channel, or a social media account.

Set a goal for your product, say, 1,000 subscribers, and be able to show a demonstrated loop of learning and experimenting you applied to reach this goal. You don't even really need to reach it. You just want to work on it consistently and record the learnings of your experiments.

If your product fails and you know why, you still achieve the goal of this exercise. You can commit to these four action items at any stage of your career to help land your dream job as a product manager.

Landing a Junior Position

As mentioned, a product manager does not have a standardized career track. The junior positions in this profession are a collection of different names you can hunt for on LinkedIn and your favorite job boards. Here is a brief overview of entry-level positions:

Associate Product Manager (APM)

Individuals in this role are already recognized as unpolished product diamonds. They have what it takes to lead a product, but due to inexperience, need support and supervision. The senior manager is there to polish the diamonds and provide APMs with various tasks to develop missing skills on assignments where mistakes can be made.

Junior Product Manager

Same as above really, however, the title is more accurate. It's rarely used though.

Product Analyst

These professionals are not officially considered junior product managers; however, people on this path are most likely to get promoted into a product management career track. These individuals analyze product and feature performance, lead product discovery and research, and look for possible product improvements based on data research. It sounds much like a product manager, doesn't it?

Product Management Intern

This role is another name for an associate/proxy product manager, but here we are looking more at a potentially unpolished diamond, rather than a confirmed one. Thus, these junior professionals are hired for a limited time and have specific tasks/projects to complete during their internship. You could call them *"junior, junior product managers."*

Product Owner (PO)

Controversial entry. If you look at a product owner's responsibilities in Scrum Guide[1], they appear to be the same as product managers. However, more often than not, a product owner will act as a junior product manager, with limited decision-making authority. The product owner will execute the *"real"* product manager's vision and strategy.

Proxy Product Manager

This type of junior product manager role serves as a liaison, representing product managers when they are unavailable. Proxy product managers have limited decision-making power. Importantly, they understand the positions and directives of product managers and can represent them when needed.

Scribe Product Manager

Well, this is simply a product manager's assistant position. By definition, the scribe's role is to write different documents for the product manager.

Ok, so how do you maximize your chances to get selected for one of these junior roles? The best advice we can give other than *"apply and interview until you are successful"* is to make sure you commit time and effort to each and every application.

Use your experience and knowledge, remember to learn from your own product, showcase your skills, and prove beyond any doubt you are a great candidate and match the requirements perfectly.

Make sure you create a custom résumé for each application you send and, if available, take advantage of attaching a cover letter.

Don't use a stiff, corporate-sounding template from the internet or ask AI to write it for you. Instead, write it on your own and convince the person reading it to give you a chance, compelled by your authentic interest in the job.

Granted, it will take time and might seem like a wasted effort when rejection inevitably arrives. That being said, you will at least stand out as someone who went above and beyond to be noticed.

Another thing you should do is ask for rejection feedback. Use that feedback to polish your résumé and know what to focus on in your preparation and future interviews.

It's important to understand that true junior positions in product management are few and far between. That means countless dreamers are looking to land limited job offers. While we don't want to discourage you from applying for junior roles, you ought to consider the more reliable path to becoming a product manager

Landing a Promotion

While a product manager doesn't typically have a junior position posted, many entry-level roles support product managers and are much more realistic to land.

This includes data analytics, code developers, designers, quality assurance engineers, project managers, customer support agents, and sales engineers. Most likely, the lion's share of IT-related jobs in companies that have product managers would put you on the right path.

Keep in mind, you will want to work as close to a product manager as possible. That means you want to become a member of the product development team. From there, you will be well-positioned to take the first step. Make your intention to become a product manager known to your supervisor. Be mindful there are only two outcomes to this conversation.

> 1. You are in an organization that supports internal career development. If so, your manager will develop a plan for you to achieve benchmarks towards your goal so you can earn your promotion.

2. You will be told, *"No, that's not going to happen."*
In this scenario, it's time to find a more supportive
organization.

Let's assume you *(eventually)* hear a *"Yes."* No one will give
you the position then and there. You are looking at months
of hard work to fulfill the development plan. That means you
will need to work even more than you're already committed
to and then some.

On top of being a high performer in your current job and
meeting development plan goals, you need to show initiative.
Prove yourself and your product skills by becoming the right
hand of the sitting product manager.

Help with research, experiment analysis, and during the
refinements. Offer to replace or represent the product managers when they are sick or unavailable. This shows initiative
and competence. It will demonstrate you are ready to take on
and manage responsibility. When the opportunity comes, the
product manager is more likely to support your promotion.

At this point of narration, you may feel like you are already
doing everything listed above and still face an uphill battle.
We assure you, this is not an easy climb. We each know this
firsthand.

CHAPTER 5

CHALLENGES FOR PRODUCT MANAGERS TO OVERCOME

Product managers dream about working on a product they love, or working for the most coveted of companies that produce top-notch products, including companies with household names, such as MAMAA.

As referenced in Chapter 3, this acronym refers to Meta, Alphabet, Microsoft, Apple, and Amazon. However, many other highly profitable and coveted product roles exist for the thousands of startups created each year all over the world. The sky's the limit with products waiting for a driven product manager to bring them to IPO – that is if you're lucky to find the right match and land a job.

Three of the biggest challenges product managers face in their current roles or upon entering the job market include getting their preferred working style *(remote, hybrid, in-person)*, landing a competitive salary, and avoiding layoffs.

Challenge #1: Remote, Hybrid or In-Person

While remote work was around prior to the COVID-19 pandemic, the number of remote roles that opened from 2020-2023 sky-rocketed.

It seemed remote work would last forever, until companies started mandating employees physically return to the office. While not every company made this change, some companies are operating with a combination of in-person, hybrid, and remote roles.

The flexibility offers appealing advantages and disadvantages. The downside is these product roles once had less competition and were limited to local candidates. Now, employers are able to cast a wider net to find specific expertise and skill sets. They may find job candidates from anywhere in the world, and perhaps even at a lower salary.

While this is beneficial for companies, job seekers also have clear advantages that put them in the driver's seat. Product managers don't have to limit their search to local businesses and can apply for a dream job with a company or product they are passionate about.

However, if you're looking for product leadership roles, the market isn't as easy. There are far fewer remote roles available for product leaders because the pressure and responsibilities are much higher than those shouldered by a product manager. Companies expect product leaders to be present and visible, and they are paying a much higher salary to have that expectation. Even if the role has a hybrid work format, you will be expected to live locally for in-person days in the office.

Challenge #2: Competitive Salary

Salaries for product roles vary across industry, location, and title. The salary and job market for product managers was hot in 2022-2023 with some of the highest compensation packages recorded on Levels.fyi.[1]

This platform helps technology professionals compare compensation packages for job roles around the world, anonymously. Product manager roles reaching up to $1 million for sign-on, base, bonus, and stock leveled out in 2024.

According to Salary.com, in 2024 the median salary for a product manager in the U.S. is $150,000, ranging from $121,901 to $178,140.[2] If you work on the West or East Coasts, your total compensation package reaches closer to $300,000 for a base product manager role, including salary, stock, and bonus.

Total compensation packages for large, profitable software companies seem to be holding steady. Salaries in 2024 for a product manager at LinkedIn range from $297,000 for total compensation, including base, stock, and bonus, to $887,000 for a Director.[3]

Challenge #3: Tech Layoffs

The biggest challenge may not be landing one of these coveted roles, but keeping your job. It doesn't take much of a scroll through LinkedIn to see the massive tech layoffs following the hiring surge in 2021-2023 while companies ramped up for growth.

Many tech companies experienced rapid growth and over-hired during the height of the COVID-19 pandemic to prepare for the surge in demand for digital services and jump in remote work. When that growth didn't happen for some tech companies, they had to down-size with layoffs.

In addition, some roles are now being replaced with advances in automation, including bots that perform tasks and AI tools like ChatGPT.[4]

CHAPTER 6

WHAT DOES IT TAKE TO BE A SUCCESSFUL PRODUCT MANAGER TODAY?

Even though there isn't one single path into product management, a commonly held perception was certain types of individuals made successful product managers, i.e., they graduated from a specific school, worked at a specific set of companies, acted a certain way, and were based in certain locations. More often than not, the successful product manager leveraged expertise in hard skills, such as having a technical background or experience in one facet of product management or industry.

Having these designated characteristics and singular expertise would lead to a promotion and subsequent rise up the product management ranks.

To be a successful product manager today requires going beyond the hard skills, such as domain knowledge and technical competencies. To engage effectively with a diverse mix of customers, colleagues, and stakeholders, one must be able to call upon soft skills,[1] defined as empathy, creativity, and problem-solving by the Harvard Business Review, to work alongside a range of individuals effectively.

This approach may surprise people who perceive a product manager's role as a mini-CEO. That perception is outdated and likely originated because one product leader historically was assigned to and responsible for a single effort or opportunity and directed a corresponding team of designers and engineers.

In our post-pandemic world, the workplace keeps changing at a rapid pace. Today successful product managers help their team shine – from direct reports to other product managers, and cross-functional colleagues – by working across a range of opportunities, underscoring the adage *"we go farther together."*

Fortunately, the route to becoming a stand-out product manager has evolved. The narrow range of styles and traits no longer guarantee a pathway to success. Instead, as the plethora of products and ready access to technology have multiplied and diversified, a growing demand for a wider range of product expertise and characteristics necessitates diversity of thought and leadership styles.

Plus the growth of accelerators, bootcamps, communities, newsletters, podcasts, and online learning offerings has opened up a wealth of resources for one to call upon and build up knowledge. Given the ready access to product information, even though hard skills remain important, it's the soft skills and focus on learning that separate effective product managers and help drive their success.

What are Poor Behaviors to Stay Away from...and Why?

Just as you can train a large language model *(LLM)*, we believe in the power of continually training our minds through learning.

For example, by learning about and coming to appreciate the value of diverse perspectives, product managers can help their team to start thinking differently.

We recommend product managers steer their brain away from outdated behaviors and instead lean into the following five areas.

Five Behaviors to Practice as a Product Manager

1. Focus on Collaboration Over Self-Promotion

For those who view getting credit as the only way to get a promotion, remember this principle: being a product manager is a team sport. By encouraging others to contribute, the product manager will benefit from the so-called IKEA effect. Consider the attachment and sense of pride many people have for IKEA furniture after they assemble it themselves. Imagine the buy-in when product managers involve team members to collaborate in the decision-making process. It leads to a wider range of ideas, potential opportunities and commitment for a successful outcome when a path forward is selected. Team members will care more about the success of the project, and potentially the company, as a result.

2. Encourage All Voices not Just Your Own

Instead of being the only person consistently speaking up in meetings, provide opportunities for others to chime in by asking open-ended questions. Even though, at the start, it may take more effort for the team to get to a decision, the level of knowledge for all team members will rise. Decision-making will accelerate, and likely be even stronger, as a broader range of perspectives and multiple viewpoints are shared and considered. Remember the goal is not to manipulate colleagues to make a decision in the product manager's favor. Instead through inquiry, the product manager is cultivating a culture of collaboration, encouraging learning, and motivating team members to reach a decision they will be more committed to achieve because they were involved in the process. Remember the prior reference to the IKEA effect.

3. Promote Information Sharing instead of Silo Mentality

One person retaining all the information and making all the decisions creates walled gardens and silo thinking. Instead, promote the power of collaboration. Use Wikipedia as an example. The online encyclopedia, maintained by a community of volunteers, was dispar-

(continued)

(continued)

-aged at the start. Now, the consensus is Wikipedia can be a launching pad for research, coupled with a critical evaluation of alternative sources. As a product manager, you should ensure your team has access to as much information as possible.

By sharing information, a product manager empowers team members to have a more thoughtful, and consequently more impactful, discussion to facilitate a productive strategic approach to solve problems.

4. Listen More Than Speaking

Depending on your interests, you may align your thinking with Jimi Hendrix, *"Knowledge speaks, and wisdom listens."*[2] Or perhaps you might consider the wisdom of Chris Conley, the founder of Joie de Vivre Hospitality and former Global Head of Hospitality and Strategy for Airbnb.

He touts owls as role models. According to Conley's analogy, owls blend into the background, so they can observe. Instead of constantly seeking credit, take time to listen to your team.

Atlassian, the global software company, champions the art of communication among team members across projects and platforms. Its software tools foster a culture of collaboration: by listening more, you learn more and are likely to become more interested in the work and contributions of your team. Extra bonus, you'll likely see your thoughts shifting towards more strategy and less about tasks!

5. Focus on Development over Burn Out

In the *Human Side of Enterprise*, Douglas McGregor[3] emphasizes collaborative leadership, focusing on employee development and strong workplace relationships to foster a versatile and connected workforce. His approach values work-life balance and encourages employees to mentor others as they advance, thereby nurturing a supportive work environment. Yes, this method likely requires slowing down to some extent—*with intention.* That is, to slow down the craziness of constant work, not create an excuse to delay. Consider the approach portrayed in *The Fifth Discipline: The Art & Practice of the Learning Organization* by Peter Senge.[4] The senior lecturer at MIT suggests teams need to slow down to go faster. As collective knowledge increases among team members, they can concentrate their efforts and see greater impact down the road. Importantly, they still have the energy to get to the goal and even push beyond it.

You may find some leaders and individuals feel uncomfortable due to this *"new way of thinking."* Bluntly speaking, they may push back on the importance of soft skills… *"Isn't a bit of competition a good thing? Don't we all try harder when there is a friendly rivalry?"*

The answer is *"it depends."* In a take-credit culture, leaders may look the other way as it's accepted and expected that stepping on colleagues' toes is part of the race for status and promotion. Rivalry may appear friendly above the surface. If you scratch below the surface, the veneer of friendly rivalry loses its luster. You will likely find team members feel threatened and consequently are not performing at their best.

As you shed poor behaviors and allow soft skills to shine, you will start to see the light. It will help you bring out the best in all members of your team as well as yourself.

What Characteristics Should You Embrace?

In the pre-pandemic era, companies often perceived soft skills as insufficient, lacking the qualities to motivate superior performance.

This thinking resulted in missing the micro-moments—*small discoveries which provide keen insight and impact team members professionally and personally.*

When everyone was stuck at home and engaging exclusively by video calls, those small moments became nearly impossible to ignore as team members literally looked into each other's homes.

This opened the door to view aspects of each other's personal lives which previously were invisible and even considered irrelevant or secondary to *"getting the work done."* As we adapted to remote work, we also learned to put people first.

Bringing empathy and appreciation for team members strengthened the dynamics of pulling together to deliver and finish strong.

We can't turn back time. We had a window into the *"real life"* of work colleagues and came to realize compassion and collaboration aren't mutually exclusive. We have learned it is not easy to create an environment in which team members – *regardless of who or where they are* – feel comfortable speaking up and engaging in constructive discussions and decision-making.

Nor is it sufficient to simply act politely and check empathy off your to-do list with pleasantries. Rather, an effective product manager focuses on soft skills to develop a cohesive and positive culture among team members.

Doing so requires three qualities: *Empathy, Consideration, and Clarity.*

Three Qualities to Embrace as a Product Manager

1. Empathy

Being the most visible person in the office has taken a back seat to qualities appreciated in a hybrid and remote world, such as authenticity and human-centeredness.

We are seeing more types of product managers emerge who have the ability to motivate dispersed and diverse cross-functional teams.

The most successful product managers embrace a growth mindset and see each interaction as an opportunity to learn.

They relish the shift from output, *"do as I say,"* to outcomes predicated on encouraging all parties to contribute their expertise to achieve a goal.

2. Consideration

One of the most powerful skills to develop is active listening. Not only are you hearing what someone is saying, you are considering the thoughts and feelings behind the statements as well.

Active listening removes any competition from the conversation and creates a space where a two-way dialogue can occur. Active listening has three characteristics.

(continued)

(continued)

Cognitive – means paying attention to individuals and internalizing the insight they have provided.
Emotional – calls for keeping your reactions in check. Ditch eye rolls. Instead, truly seek and appreciate alternative views.
Behavioral – encourages beginning a response with a summary of what was heard to encourage recall and foster understanding, as well as convey appreciation of the viewpoint being shared.

3. Clarity

Developing and delivering clarity reinforces the sense of purpose for your team by ensuring product priorities align with the company mission, vision, and goals. Employees who report having clarity about their work priorities are 4.5 times[5] as likely to say they're happy at their current company.

Why? Clarity leads to increased dependability and purpose. Team members have clear goals, know their respective roles, and understand the purpose and positive impact of their work.

By guiding cross-functional colleagues to work together as one group instead of a patchwork of disconnected individuals or functions, successful product managers are able to harness the collective brain power of all involved, not just their own thoughts.

With more diversity of ideas, the chances of success rise. What's even more powerful, the mindset and behaviors conveyed by product managers influence how their team functions, just as much as, if not more, than how the organization treats its employees.

Whichever rung you're on in the organization, you can make an impact. The individuals you work with on a daily basis will notice how your temperament influences outcomes and productivity.

Applying the influencer model, your team members will emulate these positive traits. And this will create a domino effect, spreading to other parts of the business. Adopting positivity with every member of your team will create a ripple effect throughout the organization. Their interactions will encourage others to be positive, as well.

Your behavior can be the start of keystone habits, which over time, transforms everything.

PART II: FUTURE PROOF YOUR CAREER AS A NEXT-GEN PRODUCT MANAGER

CHAPTER 7

WHERE IS THE NEXT EVOLUTION OF PRODUCT MANAGEMENT?

The role of the product manager continues to evolve. For years, it was sufficient to be a knowledge-based, T-shaped person with key expertise in a particular area *(data + healthcare, payments + fintech, etc.)* and the desire to work alongside others *(customer success, marketing, finance, etc.).*

Today, effective product managers may be viewed as an

octopus tasked with managing customer demand and needs. As the range of products has skyrocketed, so have customer expectations. Product managers must draw upon more skill sets *(especially soft skills)* and continue learning to ensure the offerings they create reflect and exceed customer needs.

The prospects of the workforce have changed as well. Employers experienced the Great Resignation during the pandemic as workers recalibrated work-life balance.

Today, the pendulum has swung again. We are now in a period where excess hiring is being countered with layoffs. Even though workers may be holding on to their jobs for dear life, that does not mean a product manager has the unwavering commitment of team members.

Instead, direct reports and colleagues are constantly evaluating whether their work is valued, considering whether their voice is being heard, if learning has stalled, or if a chasm has developed between their individual values and those of the team or organization.

At that point, they may become demotivated. They do their work and show up in person or remotely, but their conviction has fallen away.

A successful product manager will understand the power of soft skills and provide opportunities for learning to keep team members energized.

These qualities encourage trust among cross-functional colleagues and team members to navigate uncertainty and doubt, both in person and remote working arrangements.

As companies are more accepting of team members working remote or hybrid, the workforce has become more diverse and dispersed. Diversity of location requires product managers to put in more interpersonal effort through the application of soft skills to stay connected and learn about their colleagues.

When diversity is embraced, rewards follow suit. One study

by Gartner revealed that a highly diverse environment can improve team performance by up to 30%.[1] Diversity also can lead to better decision-making and higher profitability.[2] In fact, according to McKinsey & Company, the most diverse companies outperform their less diverse peers by 36% in profitability.

No nod to the future would be complete without referencing artificial intelligence *(AI)*. We don't believe AI will make product managers irrelevant.

Instead, AI will help make being an octopus more manageable. Tasks which can be automated via AI will be removed from the product managers workflow. Product requirement document *(PRD)* drafts will be written by AI and improved by product managers.

Access to learning and insight will accelerate as AI and LLMs summarize videos, text and other modes of communication.

Why won't AI replace product managers? AI does not have product sense.

Yes, there is extensive knowledge behind a response you receive from your LLM of choice, yet we know that humans do not always do what they say, nor mean what they say.

Picking up these nuances requires a human, at least today. Customers have product sense too. They detect when a company is passionate about their products.

Jony Ive[3] explains how customers can sense the care and craftsmanship put into a product, even if they can't articulate why. Product managers who are able to deliver thoughtful, well-designed and innovative products *(not just ones defined by AI)* will continue to drive customer satisfaction and retention.

Let's follow another example of why AI won't replace product managers. Some companies outsource all of their customer research. This approach separates the product team from the insights and nuances which cannot be captured in a text write-up or provided by an AI assistant. As more companies are moving to a continuous integration/continuous delivery *(CI/CD)* model, the importance of including customers right off the bat for the product development process increases.

The further product managers and their cross-functional teams are removed from the customer, the harder it becomes to have rapid feedback loops. The likelihood of a misstep or misunderstanding increases, resulting in being off kilter when it comes to addressing a true customer need *(job to be done)*.

Instead, the ideal future is one where product managers are making product decisions every day and benefiting from direct customer input.

This does not mean building products exactly to customer specifications. It means mixing together insight into customer wants, needs and expectations with market trends, emerging technology developments, and competitor insight.

Blend that together with the company's unique product know-how and you get a winning formula for success. AI can definitely play a role, but AI does not replace the valuable insight direct customer contact provides to product managers and their cross-functional colleagues.

What's more, the walls between product management and other functions will continue to be fluid. As the pace of tech innovation and customer expectations continues to accelerate, so does the speed by which a product manager must be able to bridge the connections between the two.

Yes, some companies will merge roles and others will change titles to reinforce the necessity for collaboration. Yet, identifying ways to improve outcomes for the customer by bridging cross-functional stakeholders is here to stay for the foreseeable future.

What Can You Do to Prepare?

Continue to learn and apply soft skills. Seek out others who recognize the value a diverse range of viewpoints brings. Put aside silos. Instead, celebrate when a team member offers to step out of their lane and help out.

For example, this might happen when a product manager offers to step in and assist with a user research study or a designer codes a prototype to accelerate development.

When people are working together – not against each other – individuals are able to exchange skills and learn new ones, all contributing towards a common goal and getting work done together.

Remember product management expertise can go stale and requires topping up. New competitors and technologies continually emerge, driving the constant evolution of customer expectations.

Lean into the opportunities to expand and test your knowledge throughout the product development process. Remember to use the four stage approach: ***Discover — Define — Develop — Deliver.***

> **Discover:** Uncover customer needs through market and user research. Explore the problem space which aligns with your business. Each finding is an opportunity to gain customer empathy and grow your product sense.
>
> **Define:** Refine your knowledge by testing the riskiest hypothesis. With every new learning experience, your customer empathy builds and your route to a Minimum Viable Product *(MVP)* becomes clearer.
>
> **Develop:** Pilot early stage ideas with customers to identify and build an MVP. You are gaining additional domain knowledge and creative insights which enhances cross-functional collaboration.
>
> **Deliver:** Refine your MVP and subsequent build-outs with customer feedback and cross- functional input. Once you reach this milestone, empathy for your

customers continues to grow as you dive into further iterations and improvements.

At each stage, use your soft skills to listen and learn. You will start to identify themes others have likely ignored. Don't shy away from pursuing emerging opportunities.

The uncertainty will likely discourage your competitors from treading in untested waters. Instead take small steps, learn along the way.

Lean into your gut feelings and embrace the climb. When in doubt, keep the following in mind:

- Risk is not for the faint of heart, nor is product management.
- If someone says *"you're crazy,"* you're probably on the right track.
- Stop stressing about *"doing product right."* Instead focus on what you could do to make your product better for your customers.
- Nobody has it all figured out. If the work doesn't look exactly like a case study you read, that's ok.
- It's fine not to have all the answers. Seek input from others and celebrate the wins achieved together.

Along the way, embrace an attitude of gratitude for all team members. Ditch assigning blame or taking credit for the work others contributed.

Instead have collaboration and shout-outs become second nature to your culture. Remember, you can be the difference. You can introduce change and prepare yourself and your company for the future.

By continually learning and applying soft skills, you will shine a spotlight on your own success while bringing to light the superpowers of each member of your team.

By working together, we can navigate any unknown and go farther, more quickly and with some fun baked in too!

CHAPTER 8

PRACTICE, DON'T PREACH PRODUCT MANAGEMENT PRINCIPLES

The symbol most commonly associated with the legal system is Lady Justice,[1] an allegorical personification of the moral force in judicial systems. She is depicted as a blindfolded figure holding a set of scales and a sword.

Now, what if product management was as old as the legal system and also had its *"Lady PM"* statue reminiscent of ancient times. What would she be holding in her hands? We believe she would hold a roadmap in one hand and a backlog in the other.

Although it's difficult to depict those artifacts in the real world, both are real tools every product manager needs. Of course, don't forget user stories that encompass tasks, epics, or product requirements documents *(PRDs)*.

In this ancient analogy, our hypothetical goddess of product management would no longer look like an octopus, but Medusa.[2] But enough metaphors! Let's get into the meat and potatoes of this chapter. Let's dig into the trio of main product artifacts: *Roadmap, Backlog, and User Story.*

Managing Your Roadmap

The roadmap is a tricky beast for an agile product manager. On the one hand, we are taught to work in a dynamic world, where we experiment, adapt, and figure out the best direction to take as we get feedback from our experiments.

Having a roadmap is a clear contradiction to that approach and a relic of times when waterfall method code development was the only real way to do it.

Actually, scratch that. It's tricky because of the way business decisions are made, i.e., with quarterly predictions. Senior executives need to present a plan to the company owners and the estimation of the company's value is determined by whether the plan's goals were met or not.

In a world like this, you can't just make up product changes as you go along. That's why every quarter or so, you will need to present a convincing plan of updates coming to the product that will both be specific enough to please your managers, but also vague enough to allow you some agile wiggle room.

*Eight Guidelines for Managing
a Product Roadmap*

1. Commit to Working on Goals, not Features

Ideally, your roadmap shows how you will timebox different challenges and goals over the coming months. However, let's be realistic.

Management rarely will accept goals versus deliverables. In that case, you will need to supplement goal-setting with a set of proposed deliverables.

This approach provides a compromise that might be easier for stakeholders to digest. If this doesn't fly, an alternative approach is a feature-based roadmap.

2. Don't Go Too Granular

Remember, you want to impress and inspire your stakeholders with your roadmap. The presentation meeting will be a long one and if you dive too deep, you will baffle your audience and lose the narrative.

So, focus on the high level and offer as little details as possible. On top of that, don't offer specific dates *(if possible)*, but reference an undetermined date in the future.

Make it so everyone in the room understands what you wish to do and more or less by when. Don't provide concrete details that could be interpreted as promises.

3. Stress it's Only a Draft, not a Commitment

That's right! If you make this a commitment, it will

make it harder for you to change direction later if, say, another ChatGPT pops up out of nowhere. You, as the product manager, should always pursue the best value for your product and no roadmap should stop you from doing that.

This is the way you can potentially, sensibly marry the expected waterfall approach with your agile best practices.

In this scenario, you present a great plan that can be retired with a better plan if one comes your way. That being said, there is one more *"hack"* you can potentially use.

4. Apply the "Scotty Principle"

What's that you ask? It dates back to the original episodes of Star Trek that aired in the 1960s. In the science fiction television series, the chief engineer for the Spaceship Enterprise was Captain Montgomery *"Scotty"* Scott.[3]

Scotty would knowingly overestimate the time to fix the ship's problems. This way, if he did it faster, he would have seemed like a miracle worker! For you, however, it means you can slightly stretch the truth by booking more time for a future release than the minimum time it needs to go live. This way you gain additional time for follow-up updates based on user feedback. Note, this is not foolproof pending the skills of your development team members and whether they deliver a feature on time or take far longer than their estimates, thus eating into your buffer.

(continued)

(continued)

5. Involve the Team

Before showing the roadmap to stakeholders, show it to your team members for input. Make them contributors to the roadmap, not recipients.

They should feel proud to follow and be responsible for its execution. This way they can flag potential technical risks or other issues not on your radar screen.

They are also a great test audience to give honest and informed feedback. It's better to hear it early from them than from your management stakeholders, we can assure you!

6. Show How Your Propositions Fit Into the Wider Narrative

If you can highlight how your proposed roadmap addresses product goals, as well as fits into the product/company vision and responds to market/user needs, you will be a product management guru.

This will help solidify your decisions and prove they make sense. If they don't, there might be a problem with the underlying principles. Be mindful you will need good visual cues to show the connection using colors and/or icons. If you use too much text, the roadmap will become hard to read, will cause unnecessary confusion, and will result in stressful stakeholder questions.

7. Include Visualizations

If you have any wireframes or early designs, show them! They might change 1,000 times down the road, but showing early concepts will illustrate you are progressing with your plans.

This will build confidence that the presented outcomes will land when you say they will land. At the same time, stress this is an early visual reference and the designs might change before the final release. You are offering eye candy and nothing more.

8. Make it Easy to Access and Check the Progress

This will save you time on updates. Stakeholders will be able to check your progress on their own and it will be easier for you to create a *"check-in"* presentation.

You can do it using any roadmap software, ideally one that connects to your backlog and can generate a cloud-based roadmap visual automatically. This way you will save time on manually creating and updating the roadmap graph. *"Ain't nobody got time for that!"*

This roadmap guidance should allow you to achieve a *"win-win"* situation, enabling you to work in an agile way. Plus, stakeholders have confidence that plans are in place they can trust. However, keep in mind this roadmap will need to transition into a more specific list of tasks, known as the backlog.

Managing Your Backlog

Backlog is the collection of tasks/user stories/tickets in any given company. It can develop in two ways. It's either a functional, prioritized, and well-maintained list of assignments for the development team members to take to their sprints; or, it is a gigantic pile of *"things not going to happen,"* excluding perhaps the top five priority items.

Unfortunately, given the incoming, often inconsequential

bugs, requests from different teams and stakeholders, and ever-shifting strategy and market demands, the second scenario is far more likely.

How do you avoid this and keep your backlog tidy and functional? Here are eight tips.

Eight Guidelines for Managing a Product Backlog

1. Differentiate Between a Backlog Item and an Idea

Let's start by not putting every half-baked idea in the hopper in the first place! It's ok to have a notebook, FigJam, Miro or Mural, where you collect all ideas and requests.

However, the backlog should only contain items you aim to work on FOR REAL within the next quarter or two. This way the backlog can resemble a plan, not a Christmas gift wishlist in March.

2. Set a Maximum Number of Tickets in the Backlog

The agile scrum guide suggests setting the limit at 50 backlog items. However, in my experience, only the top 20–30 tickets will actually have any chance to ever be closed as completed.

There are too many new directions, opportunities, and urgent tasks coming in that ultimately overrides the priority of tasks deep in the backlog

You might as well acknowledge the truth and close the items that will never happen, or at least move them to your idea space. This also will limit the number of meaningless bugs in your backlog.

3. Don't Turn Your Back-log into a "BUG-log"

Bugs are tasks like all others and often they are the easiest and cheapest way to bring additional value to the product. To that end, they need value and effort estimation just like any other ticket.

This will allow them to be correctly prioritized against any other product opportunity. If they don't make the cut, they don't make the cut, sorry.

There is no point in collecting bugs that will only transform your backlog into a rotten corpse of one. You want the opposite.

4. Keep the Tickets High-quality

If something makes it into your backlog, let it shine! Make sure to include the user story, impact hypothesis, requirements, and links to design and tracking specifications.

The tickets should be able to speak for you when you are not around. For more details on this, continue reading this chapter.

5. Have as Many Tickets Ready for Development as Possible

While this may sound impossible to some product managers, try to have three months of refined items ready to go. While it will be a challenge to achieve it initially, it's worth it!

With items in queue the team can dive into for the

(continued)

(continued)

next three months, you will have time to do proper long-term planning and assessment. Plus, you will have more breathing space to play around and polish potential future items.

Even if something high-priority comes out of nowhere, the *"ready for development"* items can always sit in the backlog as quality backup items that you can always take to sprint if the highest priority tickets aren't ready for development.

6. Use Visual Cues

It's much easier to look at the backlog if you can easily see a new feature task, improvement initiatives, bugs, and research.

If you add color cues to represent item status *(details needed, refinement ready, refined)* you will be able to check status at a glance. This way the backlog will become much more functional and tidy.

7. Automate Stakeholder Updates

A personal update via email can keep stakeholders apprised on specific backlog tickets. Automated status updates keep relevant people in the loop with no time investment on your end.

This way you can cover the basics in case there are slight modifications to the initiative. It keeps stakeholders updated with opportunity to comment.

8. Create a Task Document Associated with a Backlog Item

This is basically an extended version of the ticket, where you can collect all the discovery and post-development results and observations.

Collecting this information and keeping it in one place saves you hours when it comes to writing progress updates and presentations. At the same time, your tickets remain clean and hold only the relevant information.

Thus, the backlog is primarily for development and developers. Your dedicated task document is where everything you, the product manager, can store items.

Hopefully, after applying the advice above, your backlog will finally become a reliable tool, not a collection of broken promises! Now, let's move on to user stories, what the backlog holds.

Creating User Stories

For clarification, a user story in this context is a name given to a backlog's task and widely associated with companies that manage products with an agile *(vs. waterfall)* mindest. As all companies and products are different, let's change the formula for this section of the chapter. Rather than listing bits of advice, let's look at an effective format for a great user story for a new product or function in eight steps.

Eight Guidelines for Creating User Stories

1. The Title
The title ought to be like a funny joke. If it needs

(continued)

(continued)

explanation, it's not good. Jokes aside, choose a clear name that effortlessly conveys what you wish to change in the product.

2. User Story

Let's consider the agile framework mentioned earlier. It goes like this: ***"As [a user persona], I want [to perform this action] so that [I can accomplish this goal]."***
While this sentence may seem cliche, it has important functions. It makes the task at hand relatable, shows user perspective, and highlights the problem being solved.

Very often product managers design a solution and look for a problem. This framework can help prevent that from happening.

3. Task

This is effectively the same as the user story, but written in plain English. It uses simplified language to say what you wish to actually change in the product.

4. Hypothesis

You can merge this with the task section. Here you need to focus on what will change from the user's perspective and, essentially, why bother in the first place.

5. Discovery Results

If you collect all the core figures that led you to decide to invest in the change in the first place, it will be easier to communicate and convey your motivations later. Trust me, the best numbers have a tendency to get misplaced or misremembered. Of course, limit yourself to a summary. Sharing the full results can be housed in the initiative's document, as mentioned in the backlog section.

6. Impact Hypothesis

Here, you need to select the metrics you presume will be impacted and your best guesstimates about how much change in those metrics to expect.

7. Designs and Tracking Specification

Document all the links to relevant design documents. Having them handy avoids headaches and wasted time later.

8. Requirements

Writing the right requirements could be their own chapter. In a nutshell, make sure you cover all the potential and realistic scenarios: happy path, different errors, and likely edge cases.

Essentially, this could alternatively be a testing plan! Be sure to open this section for discussion. You are just one product manager and your team will likely have a ton of questions that you can answer with the right set of requirements.

Perhaps now you are able to imagine your face on the hypothetical ancient statue, juggling your responsibilities with the assurance of a prepared product manager. Hold your head high, grip your roadmap, hoist your backlog and clutch your user stories with pride.

AUTHOR REFLECTIONS: DR. BART JAWORSKI

My professional career started when I landed a remote job during college. Back in 2007, that was quite an anomaly. I started as an engineer, followed by a stint as a client support specialist, and finally as a project manager.

It was a dream come true.

One day my job title actually changed to **"*product manager.*"** This was a promotion in name only, but it happened nevertheless.

Truth be told, my responsibilities weren't those of a true product manager. I had some product decision elements added on top of my existing role, but that was about it. For a time it was good… until it wasn't.

This is when my three-year struggle to become a true product manager began.

How I broke into Product Management

I vividly remember my first job interview for a new product manager position. Suffice it to say, I was made aware that although I appeared to have the right talent and guts, I missed a lot of key required experience, for example:

- No A/B testing
- No metric focus
- No data-analysis skills
- No product strategy work
- Hardly any leadership evidence
- Didn't even know what Objective and Key Results *(OKRs)* were
- No Agile Scrum experience whatsoever

I softly surrendered.

I decided to complete my doctorate and spend the next year in formal education to get hired as a product manager. I also tried incorporating the learnings I gathered into my day-to-day work, to showcase the skills I was missing on the job interview.

I started again one year later. For two years I was rejected time and time again, even though I already had the product manager title on my résumé. But you know what? I learned something from every single failed interview.

With each rejection, I requested feedback and when I got it, I became slightly better. As time moved on, I was getting further and further into the interview process.

I even rejected an offer because it paid less and required a relocation. But after months of humiliating job interviews *(with even my loved ones telling me to quit)* I was finally getting somewhere.

Then it happened, I landed at the biggest job board in Europe, Stepstone. Despite still lacking proper experience as a product manager, my interviews were going strong due to my vast knowledge of job boards and top recruitment companies.

But then, one simple sentence blew my interview: *"I like to have control over my backlog and be fully responsible for it."* The senior product manager didn't like that answer. He was looking for someone to enact his plans. But it was not yet over.

This person recommended me to someone else. He said there's this *"one guy in the interview process"* that he might find interesting. That was it. That was my break.

This was the day I really became a product manager. With all that being said, I hope you find inspiration from my story and take away tidbits from this book that will help you land your dream job.

Treat your developing product career as your product.

Be persistent in learning, experimenting, and trying again with new insights until you are officially hired as a product manager!

CHAPTER 9

BE A PRODUCT CHAMPION, NOT JUST AN ORDER TAKER

We have learned this lesson firsthand. For us, *"launching people"* is just as important as *"launching products."* Why? When product people feel supported, their product efforts shine. We put our team members first, create a safe space and ensure they have opportunities to shine. We treat our team as people first and foremost, as we would expect and like to be treated.

Start with Empowerment

Empowerent starts with leadership, specifically a type of leadership mindset. Some leaders focus only on profit, yet enable their teams to chart the path towards revenue.

Some leaders believe the product is the answer and trust their teams to determine how to reach the desired outcome and create superior customer delight.

Each of these leadership depictions offer their respective teams a way to contribute to the definition of success – they empower their teams to make decisions, experiment, learn, and even fail.

What is missing from the mix is the focus on people. When people rank below profit, product, or process in the pecking order, true empowerment is uneven at best.

When leaders put people first and place trust in their opinions and efforts, then true empowerment occurs. We have worked for leaders who believed they were empowering their teams, yet were actually following command and control methods or micromanaging.

For example, they would issue strict requirements or guidelines which forced their team to work within narrow guardrails that prevented experimentation or innovation

These boundaries prevented their teams from exploring two-way door decisions where an easy reverse of the approach could be done.

We've also worked with managers who believe they are empowering their teams by giving them the *"freedom"* to explore new opportunities, yet question every step of the way and require extensive process updates to move from one milestone to the next.

These leadership styles illustrate micromanagement and a

lack of trust — *not empowerment.*

Effective leaders kick micromanagement to the curb. We recommend establishing a culture of empowerment through learning.

Why learning? Employees who feel they have significant growth are also the happiest.

Forbes shares, *"Nearly 80% of employees who reported they had significant personal growth also reported they were happy in their current role."*[1] As self-reported employee growth declined, so did happiness.

Besides being beneficial to the individual, learning also provides a path towards continuous improvement for the entire product group *(and organization).* Or to phrase it slightly differently, continuous improvement requires a commitment to learning.

Think about where you work today or previously. How can an organization improve without first learning something new?

Doing something differently than your competitors or prior offerings requires seeing the world in a new way and acting accordingly. In the absence of learning, companies — *and individuals* — simply repeat old practices.

Building on these statements, a culture of empowerment encourages employees to create, acquire, and transfer knowledge and to be able to modify their approach based on the new knowledge and insights.

Empowerment requires enabling teams to identify and explore new ideas, for mistakes to take place and for learning to take root. This will facilitate positive outcomes, not the delivery of preconceived, set-in-stone, unchangeable features.

How has Empowerment Changed?

When we look back over history, the cost of building a product — *and even a company* — has declined. Today, with AI, it's the computing power needed to scale that racks up the costs, yet the models themselves are available and moving more towards open source access.

Yes, the AI expertise is expensive, yet even that will come down in cost as more people learn and gain experience. But there was a time, before the cloud, when creating a product was prohibitively expensive.

You may remember or heard of companies doing one release per year. It was expensive to run an experiment, build a prototype, recruit users, deploy software, maintain the software, and everything in between.

Making a mistake could put a company out of business as the effort to iterate and try a new path would take too long and cost too much to even consider.

It was the era of the big bang releases — *before agile and lean* — when waterfall was the approach of the day.

In such scenarios, product managers were excited when they were invited into the room where decisions were made. They were told what output to deliver, and the features were defined year*(s)* in advance.

If they were lucky, they were able to have a voice to share research that had taken you months or years to gather and prepare. Around these times, companies saw engineers as the only influencers, because they knew what magic was possible with the code.

If customers didn't understand the end result, it was the customer's fault. The user experience was not the first consideration. It was a product manager's job to keep the coders moving forward.

The business model changed with advances in technology. The cloud made developing software less expensive. More and more products were introduced into the market and competition intensified. Customer input was considered right from the start and continuous x3 *(discovery, integration, and development/deployment)* became the norm. When customers didn't understand how to use a product instinctively, they simply went elsewhere.

User experience stood out, made products shine, and customers smile. This is the world we live in today.

With this framework and our rapidly changing world in mind, empowerment means giving teams the trust and autonomy to define the path forward, opportunity to constantly try out the new, and to feel safe to apply the learnings *(good/bad)* to deliver outcomes which retain and grow the customer base.

They do so collaboratively with designers, engineers, and others based on an organization's hierarchy.

Being empowered and given the authority to explore the unknown without constant management intervention is exciting and scary. Leaders spouting the importance of failure is not enough to let innovation occur.

They also need to back it up with action. Did they encourage their team members to try again, to brush off their discouragement and find a new path to success? If an individual tries something and it doesn't work, his or her career should not experience a setback.

Instead of only promoting people because of a successful product launch, factor in the experiments + learning + sharing of insight. What did they learn from the experience?

Did they share the learning? As a result, did more — *potentially bigger* — wins occur elsewhere in the organization? That's the face of empowerment today.

How Do You Create an Empowered Team?

Today the mindset and actions of employees contribute at least as much to the company's culture, if not more, than the founders themselves.

Companies have to rethink how they organize, manage, and lead — *how they empower people* — for their businesses to stand out and merit 100% of people's work energy, not the 60% they contribute to just get by until a new job is available. Increase engagement by ensuring your product culture provides purpose and belonging.

Purpose means individuals have visibility into the meaningful nature of their work and truly know the impacts are valued. Belonging aligns with psychological safety. Do individuals feel they fit in, can they be themselves, and do they feel part of a team?

Lack of purpose and engagement leads to disempowerment and lowers performance. For example, consider what success actually means at your organization.

If you only promote people who *"succeed,"* does it mean people practice *"loss aversion,"* i.e., avoiding risk because if they *"lose"* will they also lose a promotion? In such environments, it is often considered *"ok"* to sabotage or put down others because they tried a new approach which did not succeed.

That doesn't sound healthy or empowering. If this management style reminds you of your organization, it could be time to rethink how you promote or acknowledge wins, because you are not empowering people to be their best.

The question then becomes, where do you draw the line? Where does the leader stop and employee empowerment begin? Yes, there should be fluidity as irreversible decisions require an extra level of rigor than those which can be stepped back. If you approach each opportunity with a learning or growth mindset, an empowered product team appreciates the importance of continuous and rapid experimentation, such as testing and learning. They are able to make mistakes to learn while acknowledging they need to correct them quickly to mitigate the risks.

Experimentation is encouraged as long as it is done systematically through the following steps:

- Relying on *"hypothesis testing,"* rather than guess-work, for diagnosing problems
- Insisting on data, rather than assumptions, as the background for decision making
- Pushing beyond the obvious, even when conventional wisdom says it is unnecessary
- Using the above points to better manage experiments and preemptively catch risks

If these practices are followed, leaders gain trust in their employees to tackle problems under their own accord as long as they are showing learning focused on the desired outcome.

Note I didn't say *"progress"* as sometimes the learning shows the desired outcome is not realistic. This doesn't mean employees go into a black hole and magically emerge with a eureka moment.

They are empowered to speak up and share along the way —*the good, the bad, the ups, the downs.* Trust only works if it is a two-way street — between leaders and employees and vice versa.

Giving people the permission to experiment, learn, and grow should be energizing and, you guessed it, empowering!

Exercises to Encourage Empowerment

It's likely we've all sat through retrospectives before. A retrospective is a meeting dedicated to discussing what went well and what can be improved. Often used at the end of a sprint or once a product is released, a retrospective also can be used to evaluate other product activities — *such as where and how to improve empowerment.*

For example, teams often feel they are not empowered because there are too many meetings and/or leadership checkpoints.

In such situations, we've coached teams to hold a meeting retrospective.

Often the sync meetings are the first to be removed as the updates can effectively be shared over Slack, slides or email. Stakeholder meetings where discussion rarely occurs and are used to convey updates are often eliminated as well. In their place, once-a-month versions where the focus is on collaboration and communication are preferred.

Meetings that empower teams by encouraging the transfer of knowledge through learning from past experiences and from each other typically stand the test of time. We recommend exploring the following: ***Hour of Learning and Product Research and Reviews.***

Two Exercises that Encourage Empowerment

1. Hour of Learning

Definition: Effective way to establish a learning culture by ensuring your product team continues to learn by setting aside an hour to focus specifically on learning new things.

Frequency: The recommendation is to hold the hour of learning on a Friday as a way to wrap up the week. An hour of learning can happen once a month or the frequency which enables learning to be infused into your product team's culture.

Topics: It could be something a product team member has learned recently *(top themes from a recent conference)*; something you feel the team should know *(best practices for user research, difference between leading and lagging metrics, etc.)*; or guidance from a subject matter expert *(customer panel, guest from product marketing, sales, etc.).*

Prerequisites: Limit any knowledge required beforehand. Ideally no pre-work is required to ensure the hour of learning does not create an additional burden on the product team *(except for the presenter who must prepare)*.

Approach: The hour of learning structure can be diverse *(people presenting from slides, live demos, and/or interactive group work)*. Sync with the presenter to think about the knowledge being shared and the most effective way for a group of people to learn it.

Sharing: By making assets *(recordings, slides, etc.)* available afterwards, people who missed the session can refer to them later or anyone in the product team can revisit the content when/if needed.

2. Product Research and Reviews

Definition: Opportunity for the entire product group to gain insight and share feedback by reviewing customer or market research findings and demo capabilities underway.

Frequency: The recommendation is to hold the research and review session every other week or once a month. This frequency will be informed by the size of the product team.

Topics: Often, research conducted by one part of the product group informs activities other parts of the product team are exploring due to common customer characteristics and interconnected back-end experiences. In a similar manner, demoing and discussing product capabilities together as a team breaks down silos and enables learning about all parts of the product to contribute to knowledge.

Under either of these examples, you can weave in experimentation — *specifically efforts which involve searching for, and testing of, new knowledge and approaches.* Why? The most successful teams also are the ones failing the most.

It does sound counterintuitive but, to identify the optimal way to make progress towards the company's goals, these teams also are running experiment after experiment, and sharing the results along the way.

Even when experiments fail, the learning informs subsequent successes which delight customers and consequently pay the bills as mistakes are shared, empowering the entire team to build on what worked and what didn't.

If your team is not experimenting enough or at all, it could be due to some or all of the following reasons.

> **Lack of Trust:** Leadership feels only individuals at a certain level are sufficiently knowledgeable to make decisions.
>
> **Shifting Priorities:** Changes or ambiguity in company strategy, priorities, or goals leads teams to question the direction or action of teammates and leadership.
>
> **Cross-Functional Conflict:** Competitive culture encourages teams or individuals to work against each other instead of collaboratively together.
>
> **Lack of Clarity:** Vague communication and requests without context leading to misunderstandings, confusion, and wasted time.
>
> **Changing Milestones:** Shifting goal lines force teams to juggle timelines and cut corners due to unforeseen asks for more, more, and more.

If any of these scenarios sound familiar, it's time to make noise — *not make do.* The goal is to create an environment built on empowerment where people feel they can bring forward obstacles preventing solutions.

You're not encouraging individuals to simply complain about everything or act in an unprofessional manner. Instead the desire is to bring the team together to work through critical pain points.

The following is an exercise we recommend. You can do it by sharing names or anonymously. You'll know best based on the status of your team or company culture:

> **First:** Begin by asking people to write out the big issues that no one likes to talk about.
>
> **Second:** Then, capture lingering issues or resentment that no matter what the cause, needs to be addressed, or else people will continue to feel a lack of empowerment.
>
> **Third:** Lastly, in the same space, identify individual frustrations which are causing people to feel stressed and not empowered.

Once the exercise is complete assess the feedback. Divide up the information into themes and brainstorm next steps to provide closure and enable change. Decide what to work on among team members, including aspects that might not be ready to be shared more broadly just yet.

For example, if your team is concerned about change, help them to see staying on the same path — *and not evolving* — will actually result in loss to customers, the business, and to themselves.

Encourage your team members to participate in addressing the items that were captured, instead of pushing back. Remind them that continuing to rely on the old ways of working while the world moves on, is likely to be even more discouraging and isolating. Then, work with the leadership on areas that require broader input.

Establishing a Community of Practice

If you've run the exercises shared above and find follow through on the actions is lacking, recall empowerment doesn't occur when you simply go after the status quo.

If your manager is too busy to accompany you on the journey to introduce empowerment, take inspiration from this quote by anthropologist, speaker, and author Margaret Mead:

— "Never doubt that a small group of thoughtful, committed citizens can change the world; indeed, it's the only thing that ever has."[2]

Consider creating a forum with product colleagues across the business and around the globe. These groups are referred to as *"communities of practice"* or *"learning networks."*

Communities of practice often focus on shared knowledge, professional networking, and common skills, evolving around a common domain or area, or created with the specific goal of introducing change — *such as increasing empowerment.*

In a 2018 study at the University of Pennsylvania, Professor Damon Centola[3] conducted experiments with online communities to determine the threshold for cultural change.

He found when at least 25% of the community members formed a *"committed minority group"* advocating for a shift in norms, change became inevitable. Our favorite part, change actually began with ONE person in the group. So how do you go from one person to a 25% committed minority? Consider the following steps:

> **Step One:** While you might need 25%, you start with one voice — *it could be yours!*
> **Step Two:** How do you go from 1 person to 25%?

The inflection point is 3.5% of people actively engaged in the change and displayed sustained involvement. Begin by identifying a couple of colleagues who have a similar mindset as yours and are seeking more empowerment to reach the 3.5%.

Step Three: Work with the 3.5% to create a shared sense of purpose. Ensure all members of the community understand how their work contributes to something that is larger, more substantial, and more ambitious than their individual activity.

Step Four: Once you have the 3.5% involved through purpose-driven methods, work together to create an artifact that defines the desired future state. You could fill out a vision board, draft a Press Release/Frequently Asked Questions *(PR FAQ)* or write an ideal-state letter to the CEO coming from the viewpoint of a thankful customer. Have members of the 3.5% talk through the artifact*(s)*. As they do so, an even more amazing picture will form — *one all community members can rally behind.*

Step Five: Use the artifact*(s)*, to expand the discussion and invite others to explore the problem space you and/or your business is addressing. Use each other as sounding boards to refine the associated hypothesis and assumptions. Keep the conversations going to reach the 25% and retain engagement by asking another question. ***"What's your desired role in creating <the change> our community is driving?"*** Based on the responses, identify those most aligned with the empowered future. Continue to co-create with them to bring the 25% tipping point within your reach.

By joining together and creating a community, individuals are more innovative and able to solve larger problems by discussing and working through them together.

Participants are happier in their jobs, because they keep growing and learning.

As product people, we cannot stand still. Technology and the world around us continues to evolve, so must we, or we face becoming irrelevant.

As the strength of the community grows, we improve the organization's *(1)* projects delivered, *(2)* skills gained, *(3)* cost savings, and *(4)* time management. In turn, you will earn the trust and willingness of leadership to support increased empowerment.

CHAPTER 10

DRIVE MORE RESULTS WITH DATA

It's been said product managers need to be data-driven and base their actions on measurable results. While this is universally true in principle, the actual reality of day-to-day work is the product development direction often has a lot to do with gut instinct and stakeholders' egos.

Imagine a situation where collecting evidence and data will be more costly than actually trying a product experiment. However, most of the time you actually face an impossible prioritization task of 10+ competing critical entries in Jira that can't be prioritized without proper data.

Thus, how does one maintain the merits of a data-driven strategy in such a world? Let's look at a few guiding principles that will help you out day to day.

Start with the Right Questions

You can have data on anything, anywhere and anytime. But looking at ALL of it is a waste of time.

You need to ask the right data questions to have the right start. Nailing the right questions and issuing the perfect product hypothesis can be difficult, but it's the best roadmap for the path forward on the project.

Unless you invest in a proper way to nail this challenge, you may see your initiative fail before it's even presented to any of the stakeholders or team.

Of course, as discovery goes on and you get feedback, you may update your initial hypothesis. The product manager needs to lead the way on this.

Asking the right question is strictly connected with being able to eventually provide a measurable answer.

Choose the Right Metrics to Follow

Even the right questions can have confusing answers. We suggest focusing on up to three metrics that will give you the answers you seek.

We can admit — *three is an arbitrary limit.* In our experience, this gives you the freedom to pick and choose, while still keeping the right focus.

Of course, that doesn't mean your eventual product experiment's dashboard can follow only three metrics. Not at all! In fact, the more, the mer… safer!

If the dashboard tracks dozens of your product metrics, then you can discover product change impact in areas you didn't expect. The limit of three applies mostly to communication and documentation. With that, it will be easier to get all the stakeholders on the same page and follow your narration. It also will result in easy-to-understand slides in future presentations. But this will only happen if you stick to the following guidance.

Figure Out the Right Way to Observe Them

What we mean here is to make sure you can easily observe the data once it starts coming in. Sometimes the optimal visualization will be a table, bar chart, or a funnel.

You may need a dynamically updated Excel sheet. The goal here is to have the data speak for itself in a language that even the most unplugged stakeholders will understand.

To achieve that, make sure your chart's axis is labeled, metrics are easy to understand, and you do not overlay different charts on one another when their axis legend isn't really compatible.

In any case, even the best visualization will be broken if the data coming in isn't stellar.

Filter Out the Noise

You will need to work closely with your data team to understand the margin of error and any data faults and caveats that may obstruct the view.

A good rule of thumb to follow is that any metric change <5% can be a statistical error and isn't worth bragging about. Perhaps there are products and teams with stellar tracking abilities for which 0.2% of the change in the results is statistically significant and translates to millions of dollars of additional income.

If so, that is admirable. Most companies we've worked with had dozens of known data issues *(i.e., assigning a small number of A/B test users into both groups)* that made such small gains highly unreliable.

Don't Rush the Conclusions

If something is too good to be true, it probably is. Make sure any data change you observe is correct and replicable later before making a huge mistake.

Better to delay the positive news a little, rather than back down from it later. The same goes for unexpected, spectacular failures or observing no change at all — *while both can be true, we observed more often than not that such outcomes could be quickly attributed to a bug or incorrect setup of the experiment.*

Keep a good record of the numbers and findings. Dashboards and data files come and go, unless properly saved and recorded. Be the data archivist!

Dr. Bart still actively uses experimental data from four years ago at The Stepstone Group, even with a three-year break from his employment at the company. This is only possible due to his meticulous documentation process, since all the tickets and tracking dashboards from that time frame are long gone for various reasons.

Putting the results in a single document has an additional benefit. When preparing a report or a presentation, you already have the results written and formatted in a single place. With such a document at your fingertips, preparing reports for stakeholders will be a breeze.

Simplify the Message in Your Communication

No one needs to hear about the complicated details behind your data in a 30-page report. What matters most is the answer to the question you originally posed, and perhaps some unique, unexpected outcomes.

You should have all the details and answers in case someone asks. But if the email is too long, the report too detailed or the slide too crowded, the key takeaway message may not resonate with your stakeholders.

Or even worse, it might lead to confusing, irrelevant questions that derail you.

Make sure that whatever you push to stakeholders, even a layman could readily understand. These principles provide a good foundation to start being data-driven and work effectively with data obtained during your product experiments.

However, being data-driven is not limited to reporting. It also applies much earlier, when you need to obtain the data to properly prioritize your backlog and decide on your next product updates.

How to Get the Best Data

Looks for Similar Ideas Inside Your Company

When embarking on a new project, you might not be the first person in your company to think of your *(team's)* idea. Perhaps something identical or similar was already tried and tested in a different area of your product or product line.

This presents an invaluable opportunity to leverage existing data, documentation, and the rationale behind previous initiatives.

Such resources provide a foundation of well-grounded assumptions from which to build. Provided you see the relevant similarity, talk with the product manager who led the prior project.

This conversation can yield additional, undocumented knowledge, providing insights that offer a comprehensive overview of what to expect and for which you can prepare. For example, Dr. Bart recently revamped onboarding in his employer's Android app and utilized experiences gathered from three different teams working on similar features across the company.

Get Insights from Your Competitors and Similar Products

When looking for data to back up your claims, consider your competitors. Often, their work can *"inspire"* your roadmap with some relevant data included.

While obtaining precise data from competitors may be challenging, it's likely something was disclosed through company blogs or LinkedIn updates.

It's a delicate balance; the data presented may be skewed toward marketing rather than scientific accuracy. However, it's still better than pure gut feeling.

Of course *"borrowing"* from competitors is not the only source of great, data-founded ideas.

Researching quasi-competitors or checking in on other products that have an effective solution to a similar problem you are facing may bear fruit to advance your product.

However, tread lightly. For example, the Stories feature was a resounding success when copied from Snapchat to Instagram. However, it was a complete disaster when the feature was added to Skype.

Additionally, interviews with clients who were snatched from your competitors *(for B2B companies)* can provide qualitative insights. Take note, it's essential to treat these conversations as sources of inspiration rather than concrete data.

Turn to Science

Scientific research, often overlooked, is a great source of reliable data. They always go *(or should go)* through a rigorous peer review process that ensures the legitimacy of the data presented.

To find research on the topic that interests you, say gamification, Google Scholar is a great tool that will help you find relevant papers. Some might require a payment for access, but most likely it will still be the cheapest way to obtain data that supports your product hypothesis.

Web Search and Social Media

Ok, there is hardly anything enlightening here, especially since online blogs and articles don't necessarily merit credibility to make them reliable.

However, if you have time and are stubborn enough, you can try to contact the author of the data point to discuss more in detail and chew on their findings.

LinkedIn is a great tool for finding product managers behind different products and features.

It's a good launching pad to reach peers who may provide useful information that's not otherwise off limits from non-disclosure agreements.

Ok, admittedly — *the previous paragraph is basically a Hail Mary pass when no other data can be found that would help support your product initiative.*

But what happens when you still believe in your idea despite having nothing to support the hunch? You have a wealth of experience, a track record of success, and a gut feeling that's hard to ignore. Should you pursue it?

Going forth in such a situation might seem to contradict the principles we established earlier. It's similar to a poker player who starts with a mathematical strategy but gradually becomes overconfident as the wins pile up. Ultimately, hubris replaces cold probability calculations and one of the *"all-ins"* finally fails.

Will you fail if you proceed with your idea? Well, product management is, at its core, about managing risk. If your idea can be tested with minimal effort, why not take a calculated risk? Sometimes, it's easy enough to put together a basic minimum viable product *(MVP)* and see what happens.

Of course, in our poker analogy that would mean a willingness to look at the flop after no one raised during the pre-flop round.

On the other hand, if your idea requires massive investment and there is no way around it, there are two ways to push it forward:

1. Politics

In this scenario, build a coalition of support from your team and managers to take the risk. If it fails, you won't be the only person responsible. However, a more sensible option would be to invest in discovery.

2. Invest in Discovery

Product discovery could be its own chapter or book, but for now, let's just say that if data is nowhere to be found, you can always generate it. Conducting qualitative or quantitative research should give you some answers on whether your initiative should move forward or not.

CHAPTER 11

USING AI TOOLS TO STAY SHARP

If you're not already sharpening your product toolbox on AI and automation, get to it. While 2023 is considered the year of AI, 2024 is the year of AI startups.

The AI market is expected to reach $184 billion by the end of 2024 and $826.70 billion by 2030.[1] In 2023, there were 50,000 AI startups worldwide[2] and that number is projected to reach more than 70,000 in 2024.

If you're not already working for a company investing in *(or saying they are investing in)* AI or automation and/or using a tool that embeds it, you are behind the eight ball.

While there is a real concern that some advancements in AI tech could produce more layoffs, there are enormous benefits to the daily life of a product manager using many of these tools. Here are some of our favorites: ***ChatGPT, Grammarly, Orai, Amplitude, FigJam.***

Streamline Research with ChatGPT

Many competitors to ChatGPT[3] are already on the market, including Google Gemini or even Google's AI overview.

One of the most time-consuming tasks for product managers is conducting research, such as ideation, brainstorming for new products, industry trends, and competitive analysis.

While you must validate the data provided by ChatGPT, it's a starting point for research that allows you to think through and validate ideas and concepts as potential solutions to your customer problems.

Even if you cannot use this research word for word, it will fast-track the information you need to instill creative thinking and deliver solutions faster.

Improve Writing & Communication with Grammarly

If you've never used a digital writing assistant, now's the time to start. Grammarly[4] is a digital writing assistant that helps users improve their writing by checking grammar and spelling, as well as providing real-time suggestions on the clarity, tone and quality of your writing.

While Grammarly can be integrated with many tools, our most favorite use is for email communication. This can be particularly useful for communication with clients and users.

It also has utility for internal communications to be concise with colleagues and to avoid sending a passive-aggressive email to team members. Even better, Grammarly can help clean up any user stories you may be writing, as well.

Become a Better Speaker with Orai

Whether or not you enjoy public speaking, presenting is a key attribute of a successful product manager. You will need to present on a regular basis to clients, stakeholders or your executive team. Orai[5] is an AI-driven coaching application that helps users improve their oral communication and speaking skills.

The platform creates a personalized approach to each user's needs and goals and provides real-time feedback on speech delivery, filler words, energy, clarity and pronunciation.

Orai gives users insights and feedback for specific speaking engagements. So, you can tailor your delivery to the audience, length of presentation, and whether you are remote, in-person, seated, or standing behind a podium on a stage.

Get Predictive Analytics & Deeper Insights with Amplitude

Amplitude[6] helps product managers make data-driven decisions about their users to improve their products. If you're not already tracking when, how and where your users are using your product today, you should start.

Even better, check out Amplitude AI. This tool uses AI and machine learning to understand and optimize real-time user behavior in digital products, including user actions, product performance, user engagement and predictive modeling.

Predictive analytics is particularly useful when understanding how well a new feature will perform, or how likely users will engage through the different parts of the product. It tracks the customer journey for you and detects any anomalies in user behavior.

Create Workflows & Roadmaps with FigJam

FigJam[7] is part of the Figma suite of products, and is a particularly useful collaborative whiteboarding tool for product managers. Whether you are brainstorming, creating a customer journey, user persona, workflow or our personal favorite, a roadmap, FigJam hits all the notes.

Even better, FigJam launched an AI tool within its product to help users. FigJam AI is meant to create a visual for users from a simple search box.

This is a real time saver so users don't have to create their own content on the whiteboard. If you're using FigJam to run a design sprint or workshop, FigJam AI also can summarize notes and discussions, making it easier to create meeting actions or key takeaways.

While FigJam AI can't build your roadmap for you, its help in labeling, categorizing and formatting content allow you to focus on your creativity and decision-making as a next-gen product manager.

PART III: PRODUCT MANAGER TO PRODUCT LEADER

CHAPTER 12

SUCCESS FOR A PRODUCT MANAGER VS. PRODUCT LEADER

We're strong believers in the importance of a well-defined strategy. Your strategy creates alignment across your organization and team, not just for today — *but for tomorrow and beyond.*

Want everyone to understand the *"why?"* Establish a clear strategy. Want everyone to be heading off into different, unrelated directions? Ignore your strategy or don't have one at all. At the top of the strategy stack are the company elements: Mission, North Star, Vision, Strategy, and Goals. These are the key factors product leaders will be thinking about day in and day out.

In some cases, it's likely product leaders contributed to the creation of these organizational principles. At the next level are the product elements: ***Product Vision, Roadmap, and Goals.***

Product strategy may be included here, as well, depending on the product's connection to the corporate strategy. A product manager focuses on delivering results at this stage.

Yes, these efforts are in support of the company's targets, but the contributions will be measured for a specific product area. This chapter focuses on these and other nuances which distinguish the delivery of results for product managers versus product leaders.

How to Measure Success?

A product leader's success typically is measured by the product team's contribution to the company goals. Depending on the culture of the organization, company goals can span financial, innovation, and professional development realms.

Product leaders also have team goals upon which they are measured, such as working with their product managers to define targets at the product or portfolio level, which in turn ladder up to the company's portion of the strategy stack.

The success of individual product managers is defined by delivery of their capability, product or portfolio goals. They also are evaluated on their own individual goals which range from professional development to collaboration to implementing the company's mission and vision.

The following chart *(Figure 12.1)* breaks down the definitions of success for the product leader versus product manager roles. As product management evolves, the delineation between product leader and product manager has become fuzzy. For example, in tough economic times, even established organizations far beyond their startup days are requiring the product leader to run efforts which traditionally would have gone to an individual contributor.

	PRODUCT MANAGER	PRODUCT LEADER
APPROACH	Define what can be accomplished at the capability, product or portfolio level in support of the product team's goals.	Define what can be accomplished across the entire product team in support of the company's goals.
ALIGNMENT	Collaborate with individual, cross-disciplinary counterparts to achieve success at a capability, product or portfolio level.	Collaborate with other cross-discipline leads to align their teams towards collective success.
AUDIENCE	Deliver outcomes to the product leader as well as to the company's leadership team.	Deliver outcomes to the company's leadership team.
ADVANCEMENT	Continue to develop one's own functional and soft skills growth, while keeping in mind opportunities for colleagues to advance in their roles, as well.	Ensure all members of the product team are growing and progressing in their careers, both functional and soft skills, while also looking out for one's own development and success of the entire organization.

Figure 12.1

Product managers also are being asked to implement the strategic vision, provide development, and guide their cross-functional teams through fluid and unforeseen circumstances by dedicating time to spend one-on-one with each individual team member.

This scenario requires product managers to step up even further and take on more characteristics previously associated with product leadership. Generational nuances are marrying the roles shouldered by successful product managers and product leaders, as well.

Millennials and Gen Zers tend to place a higher value on collaborative success given their emergence during the internet era where shared platforms and social media influenced their personal and professional experiences. Growing up during a time steered by hierarchical structures resulted in Gen Xers and Baby Boomers typically favoring individual accomplishment.

Yet, they too are starting to embrace the importance of collaborative success. For example, when company and product politics breeds a competitive, rather than collaborative environment, the pathway to success for the product manager and product leader becomes precarious at best.

Why compete against product triad colleagues *(product, design, engineering)* or each other when you are part of the same team? Working together forges a pathway to success for all involved.

"Teamwork makes the dream work," author John Maxwell outlined concepts in his book about the characteristics of high-performing teams.[1] Winning at all costs is not really winning at all. Or to put it another way, achieving a *"win"* at the cost of other people who otherwise could help you and your organization be more impactful is just not worth it.

Ultimately, individuals, teams, and the organization suffer. Yet it's likely we have all crossed paths with individuals who are consumed with a desire for recognition, how they define *"success."*

With that mindset, they lose sight of the damage they are doing to themselves and to their team's culture. The end result of this scenario: *Everyone loses.*

Instead of setting a Gladiator culture predicated on competition, product managers and product leaders should embrace a culture of collaboration as an effective pathway to success.

CHAPTER 13

YOU DON'T NEED TO BE LOUD TO LEAD

Consider a situation in which a product coach visits a company onsite for an introductory meeting. Upon her arrival, the coach observes the executive *(i.e, the potential client)* wrap up a meeting with his team.

Near the end, the executive starts screaming at his team to be better and do more. The coach is taken aback.

She knew this wouldn't be a good fit, she couldn't work with an individual who thought yelling at team members fostered the right culture.

So, after the team left the room, she approached the executive and explained why she wouldn't be staying.

The executive looks shocked and says, ***"Isn't yelling how*** *you are supposed to motivate your team? That is what my prior boss did."*

The coach then asked the executive if he could try another approach.

The executive replies, *"Yes."*

So, the coach brings the meeting participants back into the room.

The executive then seeks opinions from the team. Ideas are then put forward, discussions are held, and the executive is amazed by the new ideas and opportunities presented through collaboration.

This is a simplistic scenario, but in reality, it works.

As product people, we are building something new and figuring it out as we go. By definition, being a product person means you're living in a world where no one knows the right answer, because if somebody did, the solution would have already been designed and built.

As the global marketplace continues to evolve, so do customers' interactions and expectations of products.

Even though a foundational job to be done *(JTBD)* is likely to remain, how it's addressed over time will change

To continue meeting and exceeding customer expectations, product people must feel comfortable to evolve their thinking and approaches. This may lead to feelings of imposter syndrome.

Realize you're not alone. In fact, approximately 82% of people suffer from some form of imposter syndrome.[1]

Always speak with your team as you would want to be addressed. Be the leader you wish you had throughout your career.

Encourage your team members and yourself to tackle unknowns together, sharing insights as you learn and make mistakes along the way. Why? Building muscle to tackle unknowns can be the *"grit"* that creates the pearl. Grit, as defined by Angela Duckworth in *Grit: The Power of Passion and Perseverance,*[2] is *"perseverance and passion for long-term goals."*

In product management, we are constantly navigating ambiguity and exploring unknown territory.

Grit is what fuels leaders — *the people who set their sights on a goal and then work diligently alongside their team to leverage the collective intelligence to attain it.*

Leadership Styles

Do you want to be a leader for whom team members would strive to work with again in the future? If so, keep the following five characteristics in mind when you set goals and define the type of leader you want to be.

Five Leadership Characteristics to Have

1. Passion

These leaders have a passion that drives them forward in their professional advancement. Their path has not always been easy or straightforward.

Notably, their grit and positive character stand out. They want to know what makes each team member tick and explores each person's backstory. They believe empowering each team member benefits the entire organization.

2. Growth Mindset

These leaders appreciate they do not have all the answers and are eager to learn. They are curious and have a positive drive to bring people together to address any knowledge gaps.

They share opportunities for growth and appreciate

---

that employees who have time for professional development are the happiest and most motivated.

They understand in the absence of learning, companies are doomed to repeat failed practices.

3. Teamwork & Collaboration

These leaders appreciate the value of an effective and energized team. They reference the power of collaboration and don't have self-promotion on rinse and repeat.

They spend time getting to know their team members and colleagues. They prioritize listening and learning, rather than pushing through rapid change.

4. Attitude

These leaders are constructive and confident, not condescending or narcissistic. They know they're human and have areas to develop, which has informed their approach to leadership.

They speak in ways others can relate to and understand. They are straight forward, upfront, and candid.

They don't put up with back channeling and make clear they prefer collaboration, over competition, between teams.

5. Resourceful

These leaders empower their teams to navigate unknowns, encourage people to learn from mistakes and share findings with other team members.

They treat people with respect and genuine interest in their input and ideas, without placing judgment or

(continued)

(continued)

dismissing their feedback. This leadership approach keeps the lanes of communication open.

They share the vision for what they believe the future holds and the potential steps to get there, but they are open to shaking up the status quo and taking a different path to get there.

Are you curious if these leadership styles are dependent upon whether an organization is in office, hybrid, or remote?

Some leaders have the mindset that having employees 100 percent in-office is a preferred management approach. They may feel direct channels are best to manage their teams or to foster company culture when people work together in person.

That simply hasn't been our experience. Instead, throughout our careers, we have benefited from diverse and dispersed locations and leadership perspectives.

In all cases, the common denominator is that one needs to put in the effort to make any experience — *in office, hybrid, or remote* — worthwhile and identify what aspects of these approaches differ from one another, for better or for worse.

First and foremost, treat all team members equally, regardless of location. Let's be realistic, even when asked to come into the office, typically at least one team member won't be able to travel and will be video-conferencing. As a leader, put your people and their contributions — *not their in-office presenc*e — first.

Create an Inviting Space for Your Team

When there is a lot of change and uncertainty, it's natural for people to avoid taking risks to protect themselves.

Risk aversion permeates both virtual and real-world hallways and water coolers, resulting in employees being reluctant to speak up.

How do you create an inviting space that encourages individuals to speak up and go out on a limb, regardless of their location? Follow these four steps: *Clarify Roles and Responsibilities, Embrace Learning, Encourage Risk Taking and Understand Your Own Psychological Safety.*

Four Steps to Create an Inviting Space for Your Team

1. Clarify Roles and Responsibilities

Reinforce why each person's contributions are needed — *not only for some far-off future, but right now.* Approximately 40% of people are introverts,[3] meaning it may take extra encouragement for some individuals to feel comfortable bringing forward bold ideas and different perspectives.

2. Embrace Learning

Explain why people should share and learn from mistakes by demonstrating commonalities across teams. Implement practices, such as mentorship programs and employee resource groups (ERGs), to foster a more supportive and inclusive work environment.

3. Encourage Risk Taking

Notice the body language, actions, and verbalizations

(continued)

(continued)

that follow a request to take on a difficult task. Thank people — *publicly, privately, often, and with sincerity.* Keep in mind the employee's perspective. Doing something that feels potentially scary *(but benefits the organization)* is asking people to take a chance and potentially step out of their comfort zone.

4. Understand Your Own Psychological Safety

Think about when you speak up versus hold back. Doing so reveals the climate you're in. Take action if you find your own psychological safety *(i.e., your willingness to take risks)* is low as it's likely your team feels the same way, as well.

While undertaking the steps outlined above, don't forget yourself. Create a space where you can shine, as well. Take time to acknowledge and celebrate your successes, big or small.

Doing so can help build your own reputation, confidence and self-belief. Don't make yourself the bottleneck by centralizing all decisions on your shoulders, instead convey trust by sharing tasks.

Appreciate the superpowers in your team and identify responsibilities you can empower your team to accomplish instead of putting all the To Do's on your list.

Unsure where to start? Just as you would evaluate a sprint or final product release using a retrospective, the same approach can be applied to increase the well-being of your team.

Retrospectives are a great way to step back and reflect, to see what is working and perhaps more importantly, what is not working.

Use a collaborative tool to enable everyone to take part equally — *introverts and extroverts.* It's a perfect opportunity to identify small changes that can have a big impact.

Creating a space where every team member feels comfortable sharing ideas benefits individuals *(remote and in-person)*, cross-functional teams and the company, too.

Being intentional about product team culture empowers individuals to be more open and collaborative. Seeking feedback via retrospectives and other means encourages continuous improvement across the team.

In the end, your authentic character — *not the volume of your voice* — will empower your team to thrive. In today's dynamic environment, the employees give organizations a competitive advantage.

AUTHOR REFLECTIONS: DIANA STEPNER

When I started my product career at Monster in the UK, I led a team of four people. The team included two product managers, one content producer, and one analyst.

We covered all of the UK and Ireland *(UKIE)*. I was able to spend a lot of time with individual members of the team and provide assistance when and where they needed my guidance.

I was familiar with their product efforts inside and out. I also served as a member of Monster UKIE's leadership team, spoke at conferences and attended meetups. Leading a small team enabled me to split my time across all of these activities.

As I advanced in my career at Monster, my leadership role expanded to more and more countries. Eventually, I was heading up product and content for Monster Western Europe. So, instead of leading a team of four people, multiple individuals spread across a handful of countries reported to me.

I no longer had the bandwidth to know the ins and outs of each product effort, nor could I set aside time to contribute significantly to any type of project my teams were driving.

This reality forced me to prioritize my time much more strategically. The top priorities at an organizational, as well as at a team level, dictated where I needed to spend my time.

How I Approach Leading a Big vs. Small Team

As my roles and teams expanded in scope, I leaned more into an innovation strategy called Horizon Thinking developed by McKinsey in the 1990s.[1]

It calls for dividing time into three buckets: *Horizon 1, Horizon 2, and Horizon 3.* I recommend this structure when leading teams of more than four to six people.

Horizon Thinking for Leaders

Horizon 1
Focus on near-term or immediate priorities, typically zero to six months or up to one year from now.

When I led small teams, I would spend 70% of my time in Horizon 1 to ensure I was able to jump in to collaborate at any time.

Leading larger teams meant I had less capacity

(continued)

(continued)

for Horizon 1. Instead, I entrusted my direct reports to focus 70% for Horizon 1. I allocated approximately 20% to 30% of my time to near-term priorities.

Horizon 2
Focus on adjacent priorities, picking up where Horizon 1 left off and extending 18 months to 2 years into the future.

In smaller teams, I would allocate 20% of my time to Horizon 2. I wanted to ensure we had future work teed up and ready to go as our near-term projects wound down. This allowed me to balance iterations that any in-market product required.

In larger teams, Horizon 2 consumed the bulk of my time. I often would spend 50% to 60% of my time here to ensure the goals and strategy across product teams were clearly defined, both at a product team and at a leadership level.

When the team was smaller, it was easier to have a simple conversation. Larger teams required more context and refinement.

Horizon 3
Focus on longer term priorities, those that should be explored now or in six months to a year's time, but will not come to fruition for two to three years from now.

In smaller teams, Horizon 3 took approximately 10%

of my time. These were efforts for which I worked with other executives. When the business and my team were small, our focus was on near-term growth and achieving milestones associated with Horizon 1 and Horizon 2.

In larger teams, I expanded my future-facing commitment as I needed to ensure the longer term viability of a bigger team of individuals and broader areas of product development. Because I wasn't as involved in Horizon 1 and Horizon 2 work, I was able to spend up to 30% of my time in Horizon 3.

Keep in mind the amount of time you are able to spend in each Horizon will vary based on the stage of your business, team, and culture.

The breakdown shared above illustrates how to divide the buckets to accommodate the size of your team and allocate sufficient capacity to the future direction of the business too

As the size of your team and responsibilities grow, it's important to empower your team to cover more of the immediate priorities, thereby allowing you to focus on the future.

In all cases, your team needs to keep an eye on each Horizon, just at varying degrees compared to the product leader. The equation would be informed by their experiences and examples mentioned in Horizon 1, Horizon 2, and Horizon 3.

CHAPTER 14

USING YOUR LEADERSHIP STYLE TO DELIVER RESULTS

We've worked at companies where an executive's style of *"listening"* was focused on interruption. The leader would constantly interject and challenge each speaker by declaring the person's viewpoint as an *"argument."* The executive had come from an extremely competitive environment and brought the combative approach along, without considering the impact on the new organization.

Consequently this leadership style fell flat with team members who had joined *(what they thought)* was a collaborative and mission-driven organization.

Why Am I Talking (WAIT)?

To determine where your leadership style would soar instead of falling flat, listen to your team. The next time you are about to interject when someone on your team is speaking, pause and WAIT.[1]

Ask yourself: *"Why Am I Talking?"* This is a communication technique you can apply that focuses on the following four steps:

The Four Steps of Why Am I Talking?

1. Observe Without Judgment
Observe what is happening and describe the situation without judgment. *"I notice ... / I hear ... / the situation is ..."* For example, start by objectively stating your interpretation without blame. *"I've noticed that limited customer insight is shared with product when sales closes or loses a deal."*

2. Express Feelings
Identify / express your feelings. *"Then I feel ..."* For example, *"I feel concerned and a bit frustrated because this gap makes it harder for product to build value and impacts our progress."*

3. Identify Needs
Identify the need behind your feeling by clearly

(continued)

(continued)

stating the underlying driver. *"My need is ... / because I would like ... / I desire ... / I need ..."* For example, *"We need a better understanding of the challenges and feedback from sales to create more effective products and features."*

4. Requests

Make a clear, positive, actionable request by asking for specific actions that can help meet the needs. *"Would you please ... / are you willing to ...?"* For example, *"Can we set up regular meetings to gather customer insights and feedback?"* or *"Can we brainstorm together ways to address the top pain points of our most strategic enterprise B2B customers?"*

As a product leader or product manager, you can use this technique when talking with your team or cross-functional colleagues to better understand their needs and feelings, especially when seeking to be a better leader.

You can transform the current situation by approaching it with empathy and curiosity; it will be easier for you to lead your team and stakeholders to find effective product solutions.

Coach your leaders and teams to create a space for conversation — *which is different from interruption* — so they appreciate the approach to take. Understand the difference between disruption and conversation.

> **Disruption:** Interruption or disturbance that breaks the flow of the meeting. It can be intentional or unintentional and often halts the current discussion.

> **Conversation:** Interactive and purposeful exchange which involves active listening and responding, encouraging understanding, and collaboration.

As world-renowned artist Fred Wilson would say:

— "It is all about zero-sum thinking. If you think the size of the pie is fixed, then you need to grab as much of it as you can. But if you are making a pie that can grow and grow and grow, you just take a small slice and let everyone else eat... Don't be selfish. Be generous."[2]

Apply a collaborative leadership approach to help your team see the bigger picture and range of pies on offered.

Aligning Different Types of Leadership Styles

As noted earlier, generational differences in values, communication styles, and work preferences can create challenges when leading a product team with members spanning multiple generations.

There are benefits too in the form of cross-generational mentorship. Embracing flexibility and focusing on common goals and values brings out the strengths of a multi-generational team.

The objective for leaders is to leverage their team's diverse strengths to enhance collaboration and drive improved results for the organization.

They will be in a better position to do so if they factor in generational preferences to increase productivity and a team's likelihood of success.

Generational change is not just about new individuals joining the workforce; it's about cultural norms and expectations shifting.

The impact on leaders is mostly felt in the shift from Baby Boomers to Gen X. The boomers had to adapt to new technologies in adulthood after their leadership styles had been formed.

The rise of personal computing influenced Gen Xers as they entered the workforce, impacting their approach to leadership from the start.

As a result, it is not unexpected to see Gen X leaders bridging the gap between multiple generations.

They have done so throughout their professional careers and personal lives, given their role helping older team members incorporate technology at work while adopting new social norms favored by their children or younger work colleagues.

Millennials were the first to grow up with the internet, and Gen Zers have never known a world without smartphones and social media. As a result, both younger generations expect instant access to information, collaboration and a more open environment at work.

One distinct area Baby Boomers and Gen Xers are on one side of the coin and Millennials and Gen Zers are on the other side is work-life balance.

Attitudes, including towards what defines success, have evolved over the years. Gen Zers and Millennials tend to prioritize meaningful work-life balance more so than Gen Xers and Baby Boomers who were more focused on career progression and financial success.

As a result, it is important for leaders to understand their company's culture, as well as read the room of their workforce. Understand preferences to inform any push asking team members to *"do more at a higher quality with less."*

The generational composition of the employees will influence approaches that fall flat versus approaches for which team members will feel empowered to rally behind.

Establish Trust

When faced with different preferences and approaches, the best path forward is to focus on building trust across and within teams. Trust can be established between generations by focusing on the following:

Establish Common Ground: Find shared experiences, values, and interests to connect across generations.

Avoid Toxic Positivity: Be realistic, embrace change, share learnings, and stay focused on outcomes.

Model Integrity: Lead by example and demonstrate consistency between words and actions.

Invest In Relationship Building: Relate to generations as peers. Start conversations. Be honest about what you don't know, share what you do.

Embrace Differences: Acknowledge and appreciate the unique perspectives each generation brings.

If you find your team is still not coming together, speak with team members in a way that also appeals to their generation, respectively.

Baby Boomers want feedback privately, where you focus on constructive criticism and specific ways to improve. Gen Xers appreciate leaders to be direct, yet tactful with a focus on results and opportunities for growth.

Millennials benefit from feedback provided in a way that also supports growth while aligning with their values of transparency and collaboration.

Gen Zers have similar characteristics and appreciate commitment to values they have and hope you share. Notably, don't pretend — *be transparent and purpose-driven.*

CHAPTER 15

MANAGING STAKEHOLDERS AT EVERY LEVEL OF THE ORGANIZATION

Steakholders are simply people about to have a great meal *(Figure 14.1).*[1] No, wait, that's STAKEholders we are talking about. Let's start again!

Stakeholders are the lifeblood of any project, encompassing anyone and everyone who benefits from your work as the product manager, both directly and indirectly.

This includes shareholders, who have a financial stake in your success, and developers, whose bonuses may hinge on the outcomes you drive.

However, not all stakeholders carry the same weight; some may simply have a passing interest in your product updates, while others, having invested substantial amounts of money, expect to see a return on their investments

A Seat at the Table for Your Steakholders

Figure 14.1

Types of Stakeholders to Manage

Users and Clients

(continued)

(continued)

Users and clients are your primary stakeholders. Without their involvement and, more likely, money, no product could ever endure. In return, they receive the solution to the problem they hired you to solve.

Members of Management

Another obvious pick when it comes to a textbook definition of a stakeholder. Their stake is the company's success, which derives from the hard work of the teams they manage.

Development Team

The success of the product on which members of the development team work is directly reflected in their career advancements and bonuses, not to mention the sense of pride and accomplishment from their work.

Other Teams (such as Sales and Marketing)

Their vested stake is very similar to the development team. Arguably, the success of every team associated with a given product depends on its outcome in an obvious way.

The sales team needs arguments to sell the product, marketing needs to have features to highlight, etc.

Investors or Shareholders

These parties are directly involved in the company's and product's success, however, their focus is seeing a profit. Unfortunately, that doesn't always correlate with building the best product and user experience.

Suppliers and Vendors

These stakeholders often drive product development with their services but also rely financially on your product to stay in business.

Unless they already have enough clients to bankroll their businesses, your product might need to adjust to meet their requirements, not the other way around.

Communities

Very similar to users, but going slightly deeper. Specifically, communities are organizations or groups of people who depend on your product to operate.

Failure to address those groups might reflect poorly on the product sentiment and also cause a lot of harm to large groups of people.

Government Entities

Well, it would be ignorant not to include governing entities as a major stakeholder. They set the rules of the game.

If you don't play by the rules *(even when those rules change)*, you lose and it means the end of your product.

Generally speaking, most often when stakeholders are mentioned, the phrase refers to those who assign tasks and eagerly await the delivery of the product and its updates

Among these, the shareholders and budget masters typically hold the most sway, as it's your responsibility to convert their investment into profit.

To that point, we could easily paraphrase a saying about spouses: ***"Happy stakeholder, happy product manager's*** *life."*

Thus, to keep your stakeholders happy, you must engage in a set of actions that will ensure their satisfaction. Let's discuss how to make it happen.

Keeping Sponsors Happy

The golden rule to remember about businesses is they love predictability. Product managers, as steward leaders of business, need to ensure realistic and transparent planning, accompanied with a plan for following and sticking to the time period.

To this end, frequent communication is paramount to any product's success. Updates via email or a dedicated document should coincide with new releases.

Ideally, they should occur every week or two. In a fast-paced start-up environment, daily briefs may be necessary to convey the constant activity of product development.

These communications should be concise, welcoming, and informative, offering a snapshot of recent updates and a peek into what's ahead.

Maintaining an archive of these communications is beneficial, especially when preparing for comprehensive presentations on the product's status.

This archive also can serve as a resource if stakeholders feel they're out of the loop due to emails landing in spam or being filtered out. Once communication barriers are resolved, you can easily resend past updates or direct stakeholders to an online repository of all communications.

While update emails will work most of the time, significant stakeholders may require less frequent but more personal update meetings.

These don't need to be formal; they could be as casual as a lunch or a quick catch-up. Focus on reporting what stakeholders care about, avoiding mundane details like the number of bugs fixed.

When faced with uncertainties, use reassuring phrases like *"We're still looking into this"* or *"More information will be available shortly."*

Don't be frustrated if stakeholders ask about details already covered in the emails; it's common to miss an email occasionally.

Even if your communications go unread, they help you organize your thoughts and quickly retrieve information when needed.

That hopefully will lead to positive feedback. As silly as it may be, even brief acknowledgments like *"Thank you for the update,"* can be gratifying and sustain motivation and good spirits among the product team.

However, it's not only about communicating regularly and transparently. When first joining an organization, review recent emails from previous product managers to understand their style and content.

You may want to adjust your style and tools to mirror what is familiar to stakeholders, to avoid potential misunderstandings or confusion with your updates.

Of course, nothing is stopping you from introducing better communications for your organization. You may find the updates sent by your company don't meet your professional standards. So, you may want to lead efforts to make communications to stakeholders top quality.

Of course, to provide excellent updates, you first need great product updates to share. Ideas for those will come from different sources and stakeholder requests are one of them.

You may be in a situation where the *"requests"* are really commands, but hopefully, you are in a company where you can say *"No"* to your stakeholders.

This can be a challenging conversation to have for some stakeholders.

Let us help you by providing a quick rundown of how to say *"No"* the right way with eight tips:

Eight Tips on How to Say "No" to Stakeholders

1. Treat Every Request as a Point of Inspiration

As a product manager, you are on the lookout for the best opportunities to add value to your product at a minimum cost and risk.

That means you need to be certain that an idea meets those criteria and you need time to look into it.

2. Give Yourself Time to Research the Request

Only when you are sure you want to do it, put it in the backlog.

3. Be Diplomatic and Transparent in Your Communication

Every request should go through a verification process. Your stakeholder needs to be aware of that.

However, you will need to dedicate time to explain that you need your numbers and time to give a final assessment.

4. Keep Common Sense

Consider potentially great ideas and avoid things that are infeasible or simply will never happen.

Don't deceive your stakeholders, give a straight answer when you already know it is not a good way to go.

5. Take Your Time Explaining Why You said "No"

While presenting the reasons a request is rejected, don't rush it.

Explain your thinking, and show the data that proves the initiatives commanding your attention at the time are well chosen and are set to solve serious challenges. Let the stakeholder follow your way of thinking and back it up with data.

6. Be Data-driven

Whether you are saying *"No"* outright or after research, show all the figures that are associated with your answer.

This will take time and effort, for sure, but it will be less of a fuss than being pushed to work on something that does a disservice to your product.

7. Encourage Stakeholders and Team Members to Share Ideas and Requests

Some product managers are never asked to do anything out of scope because they have a reputation for rejecting any request that floats their way

However, never saying *"Yes"* is a mistake. You need to embrace creativity and initiative and surround yourself with a team of *"unofficial junior product managers"* who share your passion.

One day, they may come to you and bring that eureka moment!

8. There are No Stupid Ideas!

While something may seem stupid right now, it could turn out to be the brilliant lightbulb in the future.

(continued)

Take note of all the incoming ideas outside the backlog and revisit them from time to time. Who knows? Maybe one of them was requested ahead of its time.

These are the fundamentals of stakeholder management. Insofar as the ones that tell you, or ask you, directly what to do. Keep in mind you also need to address the needs of your users and clients.

So, how do you balance the needs of stakeholders who require steady, preferably growing income, and another audience who prefers to pay as little as possible?

Balancing the Worlds of Sponsors and Users

It is frustrating as a product manager to prioritize leadership and stakeholder expectations, instead of user needs. The textbook theory for product managers claims we primarily serve the users, followed by the business.

In reality, the bottom line and budgets of the business come first.

Be mindful you have to be very careful and picky when you work on user-centric projects. So, here are eight techniques that will allow you to provide the best quality updates while also keeping the senior stakeholders happy.

Eight Techniques to Keep Stakeholders Happy

1. Quantify the Feedback

As a data-driven product manager, you need to put a metric on user feedback and customer satisfaction ratings.

This could be user ratings, Net Promoter Score *(NPS)*, or ranking of positive to negative comments. It's up to you! Just choose a sensible metric and make sure it goes up. This will help you assess user sentiment and use it in your decision-making process and communication updates.

If users rate your product high, you most likely can focus on pleasing the sponsors with a clean conscience. On the other hand, if the product rating is terrible, you will have a better argument to address it with your senior stakeholders.

2. Streamline User Feedback

Feedback will come from different sources, which include the sales department, customer support, mobile app store ratings, and more.

Make sure they all land in a single system; or assign someone to consolidate the feedback into a single report, refreshed periodically.

This will help level the feedback from users and senior stakeholders, making it easier to show the expected value, cost, and risk associated with any potential initiative regardless of its source.

At the same time, take into account the sponsor's voices have the ear of top dogs in the company and it's your job to lend an ear to the user's voices.

(continued)

(continued)

3. Highlight the Learnings

Be the advocate for users! Use the sprint and road-map reviews to highlight the core user feedback and learnings.

With that, the people who hear what you learned from user feedback will better understand their perspective and may be more likely to favor initiatives that will please the users.

4. Find Common Ground between Users and Stakeholders

Killing two birds with one stone will work out great here. If you can find stakeholder requests that will also cover confirmed user needs, that makes you a superstar product manager.

These might be few and far between, but finding user feedback that validates senior stakeholder requests is a great way to ensure the request's validity.

6. Invest in Pre- and post-Launch Discovery

Feedback from users will likely come from those who are unhappy or dissatisfied. If you invest resources in product discovery, you can learn user sentiments before investing in development.

Also, if you perform post-release discovery, you have more chances of understanding what went well and learning from happy clients.

7. Highlight User Success Stories

If you listen to your users and help them achieve something huge, brag about it to your stakeholders, clients, and potential clients.

Let them know what you did well for your users and you can do it again. Moreover, share the data that proves it simply makes sense, and how the business side also benefited.

8. Don't Act on User Feedback Directly

As previously mentioned, feedback often will come from unhappy users. This may not represent your whole user base and greasing the squeaky wheel might not be best for everyone.

Treat feedback as inspiration for research, before it actually makes it to your backlog.

There you have it. This chapter examines stakeholders and how to manage different groups effectively.

Keep the following bit of wisdom top of mind as you thread the needle on stakeholder needs and priorities. Leave room for innovation.

Consider a quote attributed to Henry Ford:

— *"If I had asked people what they wanted, they would have said faster horses."*[2]

If you only stick to addressing user feedback and abandon brave, risky updates that could shake up the product and market, you risk relegating your product to the horse and buggy museum.

CONCLUSION

To work better together, we need to appreciate the attitudes and expectations each individual brings. The foundation for any relationship is trust.

Building trust in the workplace enhances collaboration, innovation, and productivity. Start by finding shared experiences, values, and interests to connect across generations and do so in a realistic manner that focuses on outcomes.

Toxic positivity — *believing everyone will simply get along wonderfully* — is not an authentic approach, and a turnoff to team members across all generations.

Instead, as a product manager who is working with a range of stakeholders likely representing multiple generations, lead by example and demonstrate consistency between words and actions.

Doing so reinforces the foundation of trust while acknowledging and appreciating the unique perspectives each generation brings.

What has Changed in the Workplace Culture?

When we consider dynamics that have changed in workplace culture, the biggest shift is the labor pool can span four or more generations.

As previously mentioned, generational differences in values, communication styles, and work preferences create both challenges and opportunities when leading or being a member of a product team with members from different generations.

The generational ranges and key characteristics are defined as follows:

> **Silent Generation**: Born between 1928-1945. Loyal, strong work ethic, disciplined. More traditional and cautious in embracing change compared to later generations.
>
> **Boomers:** Born between 1946-1964. Individualist, driven, competitive. Seek choice. May struggle with work-life balance and younger bosses.
>
> **Gen X:** Born between 1965-1980. Independent, cynical, pragmatic, distrustful of authority. Value personal achievement and work-life balance.
>
> **Millennials (Gen Y):** Born between 1981-1996. Tech savvy. Educated, confidently optimistic, achievement oriented. Want meaningful work and flexibility. Delaying life milestones.
>
> **Gen Z**: Born between 1997-2012. Digital natives, social-change oriented, entrepreneurial. Value diversity and inclusion. Less optimistic, more isolated.

On the positive side, research findings indicate 81% of individuals ages 18-94 want to work with different generations to improve the world. According to Cogenerate[1], the generation with the strongest interest in working across generations is Gen Z.

The survey found 76% of Gen Zers and 70% of millennial

respondents wanted more opportunities to work with other generations.

For example, providing cross-generational mentorship, embracing flexibility, and focusing on common goals and values can help bring out the strengths of a multi-generational team.

Why is this important? Leveraging diverse strengths enhances collaboration and drives better results across organizations.

Factoring in generational preferences increases productivity and a team's likelihood of success. American psychologist Jean Twenge, Ph.D. provides useful insights in *Generations: The Real Differences Between Gen Z, Millennials, Gen X, Boomers, and Silents—and What They Mean for America's Future*. She writes, ***"The era when you were born has a substantial influence on your behaviors, attitudes, values, and personality traits. In fact, when you were born has a larger effect on your personality and attitudes than the family who raised you does."***[2]

Consider the information age. Baby Boomers adapted to information technology in adulthood, whereas Gen Xers grew up during the rise of personal computing and Millennials were the first to grow up with the internet.

Members of Gen Z have never known a world without smartphones and social media. A shift towards autonomy and isolation in the Digital Age also affects social, cultural and work norms.

Millennials and Gen Zers are recognized for high levels of self awareness and prioritization, shaped significantly by digital culture. Attitudes towards what defines success have evolved, as well.

Millennials tend to value meaningful work and life balance more than Gen Xers and Baby Boomers, who focus more on career progression and financial success. When we think of work styles — *remote, in-person, hybrid* — taking into account generational preferences will maximize efficiency and output.

Even though Gen Zers love flexibility, they also appreciate structure and are often the ones requesting in-person work.

Millennials prefer flexible schedules and remote work options as they have already established their social circles and appreciate having time to explore hobbies outside work. Gen Xers are independent and want to be trusted to get their work done without micromanagement.

Many prefer opportunities to work in a remote or hybrid setting as they have family lives to juggle, as well. Baby Boomers are similar to Gen Xers in appreciating the opportunity to strike a work-life balance.

That said, they also value teamwork, collaboration and opportunities for mentoring which means they will seek out interactions with other generations.

What Considerations Need to be Made Today?

We are all strapped for time, feeling stretched to the brink trying to continually find ways to do more with less. This relentless drive is leading more individuals to shield their well-being by dialing back their work efforts.

Increasingly, the sentiment is clear. In an environment where people feel undervalued, contributing just 60% to 80% shouldn't come as a surprise. It's a reflection of how they feel treated.

Change is never easy. If it was, the word *"transformation"* would not send shivers down our spines. Instead, to help your team or yourself achieve success, focus on the first step. Just the simple act of dipping one's toe in the water to get started compels subsequent forward motion.

Change is more likely to follow. With repetition and time, progress becomes a tiny habit that picks up steam across the organization. The biggest challenge to success is not technology, it's people.

If you're facing a situation where leaders only give little nibbles of empowerment and opportunity due to a lack of time, trust, budget, or other circumstances…don't give up.

Acknowledge you are not alone.

Over 40% of workers say they are burned out. In fact, the situation has gotten to a point that mental health is a key differentiator for candidate when evaluating employers, especially for Gen Zers.

Imagine what amazing things could happen if companies provide opportunities for employees to learn new skills, develop their careers and move beyond simply being satisfied because they are earning a paycheck.

Approaches to take include moving away from the idea that the traditional career ladder is the only way to advance. Look for opportunities using a lattice or portfolio approach to achieve success.

Instead of always reaching for the next rung in the ladder, the career lattice focuses on lateral experiences that develop a broader range of skills.

Consequently, this builds more space for personal development. Stepping into an adjacent role to gain skills is an example of applying the principles of a career lattice. To decide whether a career lattice approach is right for you or for your team/organization, consider the following:

Types of skills you want to develop:
Move beyond the title and think about the capability gaps you need to fill.
Existing skills applicable to other roles:
Check your eight *"arms."* Product managers are octopuses with a range of skill sets to deploy.
ABC – Always building community:
Start building connections early, share your aspira-

tions, and explore opportunities to shadow roles or
next steps.

Not convinced by the term *"career lattice?"* The key is to
think beyond climbing the *"ladder"* that goes in one direction
to achieve success. Other career paths refer to a *"portfolio"*
approach.

Similar to the career lattice, a career portfolio is based on
being flexible and adapting to the ever-changing world around
us. In a portfolio model, expand your impact outside your
organization by enlisting with adjacent or complementary
roles. The key is to let your interests guide you, not stick to an
unwavering path.

For example, consider pursuing fractional or advisory opportunities to supplement your day job.

Top Skills to Stand Out in Today's World

In today's environment, employees are less likely to leave...but
trust us, they're looking.

When an opportunity arises that respects them as people
and recognizes unmet needs — *financial, career advancement,
mental wellness, family, etc.,* — they will leave.

Has your company seen people depart for organizations in
a similar space? Transferable sector knowledge is likely a factor, yet there is stress in the uncertainty of starting over again.

As a leader of a team or indirectly for cross-functional colleagues, consider how you can help promote and retain those
around you.

If they leave, their work could fall on you or you could be
tapped to onboard their replacement in addition to your other
day-to-day tasks.

During your next 1:1, either with your manager or cross-functional colleagues, try the following techniques to identify learning and growth opportunities for others.

They'll appreciate the consideration, skill development, and increased visibility. Plus, you'll gain experience in positive motivation. Ask the following questions: ***Are there knowledge gaps in growth strategies or objectives? Are there alternative approaches worth exploring? Are there near-term wins to rally around?***

Three Questions to Ask Your Manager
or Colleagues During 1:1s

1. Are there Knowledge Gaps in Growth Strategies or Objectives?

Do sales, marketing, finance, development, and design have the insights they need to support the company or team goals?

Was their input sought when objectives were defined? You can be the bridge by sharing context as you work collaboratively on efforts.

Ensure your cross-functional colleagues help to shape the direction of the product team's work. Doing so early on prevents late stage rework and increases the likelihood of reaching goals as everyone is aiming in the same direction.

2. Are there Alternative Approaches Worth Exploring?

Often the most powerful leaders only have the capacity to make snap decisions. Instead of embracing their own empowerment, they're fighting financial constraints or the growth clock.

If your leader has reliable instincts, that's great. However, if hasty, isolated decisions aren't ideal, collaborate with your team to quickly test reversible decisions through short experiments.

Positive outcomes can improve team visibility, spur innovation, and broaden input in future decisions.

3. Are there Near-term Wins to Rally Around?

Short term milestones are wonderful as long as they are grounded in a supporting strategy and longer-term vision with leadership buy-in. When work feels transactional, the sense of purpose often evaporates.

Hold a brainstorming session to map three-month outcomes to longer-term goals *(up to five years)*, using short-term choices that are meaningful and actionable. This approach links immediate efforts to future ambitions, motivating and developing your team while emphasizing ongoing contributions.

You might be wondering, if I pursue these steps, won't I just be creating more work for myself and my team? Won't this lead to greater dissatisfaction or set false expectations?

To be clear, if you just say *"do this,"* then yes. The suggestions listed above are not for ego gratification.

You're discussing them with your colleagues to build the most thoughtful, impactful product and career development opportunities in a way that encourages everyone's success, not just flexing short-term, transactional politics.

After applying these approaches, take time to reflect. Have your efforts to provide growth and career development for others been successful?

Celebrate the wins, even the small ones. It's likely there are areas for improvement, as we're all human and a work in progress.

If you're finding one or two colleagues are not being receptive, focus on your approach. For example, when asking — *"is there anyone who disagrees?"*— folks may perceive your language as controlling or intimidating.

Ensure your delivery is coming across as encouraging, not judging or putting them down. People also may feel overwhelmed. You can narrow your approach to the top one to three highest-priority items to create growth for colleagues and yourself.

This approach will make it easier for people to remember, be present, and contribute — *leading to greater learning and opportunities for development.*

It's crucial to remember that motivations can differ from person to person, with certain approaches resonating with some and not others based on unique needs and circumstances.

That's ok and to be expected. Respect those variances. Remember, you have them too!

What Stands Out for Product Leaders?

Keeping generational diversity in mind and determining the best way to navigate preferences to set up the entire team for growth and productivity often can be perceived as a low priority for product leaders.

Yet, brushing aside the personalities and values of team members can create an environment where individuals do not feel heard or appreciated. Ultimately, that is detrimental in all scenarios, particularly when the team is remote or hybrid. Consider that 82% of people suffer from imposter syndrome, which is a deep fear of being exposed as someone who doesn't know what they're doing at work.

Throughout our careers, we've found product people experience imposter syndrome more than other roles in the work triad.

We chalk it up to the premise that product people tend to be generalists interacting in a world full of specialists. As product people, we often are building something new and figuring it out as we go.

That can be overwhelming and lead to self-doubt: am I doing the right thing, in the right way, at the right time? And, if I'm not, how long until someone else figures it out?

To put another way, being a product person means you're living in a world where no one knows the right answer. Think about it. If somebody had the solution, the product would have already been built.

When you add in the nuances of generational differences, layered on top of remote and hybrid work options, there is a high likelihood members of your team have differing views and may not feel comfortable sharing them.

Add in the ups and downs of today's economic and uncertainty around the world, workers across the generations are feeling burnt out and wondering if their job is at risk.

That's one common denominator that's putting worker anxiety on high alert. Instead of spiraling into confusion or ignoring the situation, empower your team to conquer imposter syndrome and FUD *(fear, uncertainty, and doubt)*.

Today, in our rapidly changing product world, we are constantly trying out the new and unfamiliar as we look to bring delight to our customers. Exploring the unknown day after day is exciting and scary. Encourage your team and yourself to tackle unknowns and to get comfortable being uncomfortable.

Below are four tactics for product leaders to apply with your team. You'll be amazed at what they will accomplish.

Four Tactics for Product Leaders

1. Don't Pretend to be Perfect

Nobody's perfect and no one has all the answers. Dispel the myth that product leaders are expected to have clarity and conviction even in the face of the unknown. It's ok to say, *"we don't know yet, we are trying to figure it out."*

2. Make Courageous Bets

To take on a challenge, you don't need to have all the answers. Instead you can unpack the questions together as a product team.

Admitting to not having all the answers takes courage. It takes courage to be wrong; it takes courage to pull the plug and put aside an idea or retire a product; and, it takes courage to build a culture that encourages vulnerability to swing and miss.

You can't hit it out of the park without trying.

3. Check Your Ego

Courageous teams are curious, not cynical. They know individuals are more effective when they work together as a team.

When you embrace the struggle and uncertainty of being a product manager *(PM)* and leading a team of other PMs, you'll gain the individual and collective wisdom that incorporates courage into your team's culture. And, remember the start of the journey begins with you as the leader.

4. Believe in Your Team

When you believe something can be done — *really believe it* — your mind will find the ways to do it.

Set clear expectations and remember there is no such thing as too much communication, especially in our remote/hybrid world.

Make your expectations of *"good"* clear, then repeat, repeat, repeat. Watch the culture you have created motivate team members to turn *"good"* into *"great."*

What Stands out for Product Managers?

Product managers are typically part of cross-functional teams. With that said, product managers and product leaders share a number of commonalities in their roles and responsibilities.

Although the previous section targets product leaders, the same points can be taken to heart by product managers, too. That said, there are specific nuances for product managers to keep in mind.

For example, we often hear product managers ask if individual contributors can change the culture at their organization. The answer is *"yes."* You can change the culture at any level of the organization.

The key is to be empathetic with your customers, cross-functional teams, and with yourself. By doing so, you encourage others to learn from both wins and losses. You can be the catalyst to spark a culture of trust and respect.

Given our remote, hybrid, in-person work world, product managers need to intentionally check in with team members, no matter where they are working.

Make every individual feel welcome and appreciated as a valued contributor. Tap into self-awareness. Be attuned to taking both success and failure to heart, learn from one's own mistakes.

We would go so far to say that being self-aware separates good from great product managers. Similar to the points mentioned earlier with regard to product team leaders, product managers need to recognize it's not all about you.

Be open to feedback and invite others into the conversation. Being a product manager *(and a product leader)* is not just about getting ahead.

It is about how many people want to take the journey with you. Here are four points we recommend product managers keep in mind given the environment we are working in today:

Four Key Takeaways for Product Managers

1. Adopt Clear and Concise Communication

Product management is a complex and rapidly evolving field, and it can be difficult to keep up with the jargon and concepts.

It is essential to communicate clearly and concisely to avoid confusion and misunderstandings. Think about the times in your work life you wished for someone to just speak simply.

2. Keep Audience Diversity in Mind when Communicating

Product managers interact with a wide range of individuals. Each audience has its own unique needs and expectations, and it is important to tailor your communication accordingly.

3. Be Authentic and Transparent

Trust is essential in the collaborative world of product management. People are more likely to trust those who are honest about their own limitations and willing to admit when they don't know something.

4. Embrace Observation, Practice and Feedback

Most of what we know comes from observing other people. Once we have seen a task, practice *(not perfection)* is essential to progress. To improve, feedback is necessary. Be open to feedback to continue to improve.

Remember these points when stepping into a new situation. Think about the work and what you're seeking to accomplish — *why you're there.* Embrace your imperfections and invite others to do the same. Your unique perspective and willingness to learn will build trust and create a more innovative and impactful future for everyone.

HOW MANY STARS WOULD YOU RATE THIS BOOK?

★ ★ ★ ★ ★

Authentic reviews contribute to raising awareness of books among readers. If you got just one insight from this book, we would be so grateful if you were able to leave a review, it can be as short as you would like. You can get right to the review page by going to https://NextGenProductManagement-BookReview.com.

Thank you so much! Teresa, Bart and Diana

ABOUT
TERESA CAIN

Teresa Cain is an author, speaker, entrepreneur and technology executive who has led global product and user
experience teams for more than 17 years and has held significant
roles at SS&C Technologies, H&R Block, Lexmark, Netsmart

Technologies, Service Management Group and TreviPay.

Teresa is the Founder of Lucid Startup Consulting[1] and has diverse experience leading product management, product design, research, strategy, and innovation sessions as a coach and speaker for startups and seasoned technology firms.[2] Her impressive journey includes hundreds of high profile organizations including teams at Amazon, Microsoft, Tesla, Vanguard, Deloitte, and Red Ventures.

Teresa is the instructor of the highest-rated and best-selling courses on Udemy, *"Product Strategy & Roadmapping"* and *"2 Hour Design Sprints: Learn how to solve problems and design products in just 2 hours vs. 5 days."*[3] The first edition of her book *Solving Problems in 2 Hours: How to Brainstorm and Create Solutions with Two Hour Design Sprints* released in April 2023 and was awarded Best Seller and #1 New Release in Business Technology Innovation, Strategic Management and Market Research Business books awarded by Amazon.[4]

Teresa regularly speaks at conferences on design thinking, customer experience, and product innovation. In 2024, Teresa received a Forty Under 40 Award from Ingram's Magazine. Teresa received a prestigious Emerging Scholar Award in 2023 from the International Conference on Design Principles and Practices and presented her research *"Putting Into Practice Evolving Design Thinking Methods at Technology Firms: The Evolution To 2 Hour Design Sprints."* She was also a 2022 Women in IT Summit & Award Series Finalist for Advocate of the Year.

Teresa earned a BS in Journalism from the William Allen White School of Journalism and a BA in English from The University of Kansas. She earned an Executive MBA *(EMBA)* from Rockhurst University and a Master's of Integrated Innovation for Products and Services *(MIIPS)* at Carnegie Mellon University, focusing on product and user experience design principles. Her certifications include Pragmatic Marketing Certified III ©, Net Promoter Certified Associate, Certified Scrum Product Owner, Certified Scrum Master, Project Management Professional and Lean Six Sigma Green Belt. She also completed Northwestern University's Kellogg School of Management Executive Program for Product Strategy methods, a program with a focus on discovering, developing, managing and marketing products as a business.

ABOUT DR. BART JAWORSKI

Bart Jaworski, Ph.D. is a Senior Product Manager, Product Management mentor, writer, content creator, and academic. Bart has more than 10 years of professional experience, including his work leading the video call Skype team and Android Mobile app for Europe's biggest job board network, Stepstone Group.

He is also a well-known LinkedIn personality with more than 100,000 followers with whom he shares his product management experience daily. In addition, he authors two product management blogs on Medium[1] and LogRocket.[2]

Bart is the instructor of a highly rated, best-selling online course for product managers *"Great Product Manager: Product Management by a Big Tech's PM"* that is available on his website *(www.drbartpm.com).*[3] To date, more than 20,000 people have enrolled in this course. Hundreds have written testimonials crediting this course for helping them to land their product positions. Bart has completed many certifications during his career in product management and agile methodologies, including Certified Scrum Product Owner, Product Faculty's Crystal Training™ for Advanced PM, and Product Faculty's AI Product Management.

Bart earned his doctoral degree from the Technical University of Gdańsk in the field of Automation Engineering with his thesis titled *"Evolutionary path planning system for maritime objects in high-density traffic areas."* He has authored or co-authored several scientific papers on similar themes.

ABOUT DIANA STEPNER

Diana Stepner is a seasoned product leader, coach, mentor, and technology executive with over 20 years of experience in product management, innovation, and user experience.[1] A Silicon Valley native, Diana has led diverse, global digital teams across startups, enterprise organizations, digital agencies, and software consulting. Her impressive journey includes significant roles at Epiphany, Cheapflights *(acquired by Kayak)*, SimplePractice, Salesforce, Monster, Pearson, Razorfish *(part of Publicis)*, and Epsilon. Diana is the Co-Founder of Product Gold, where she focuses on product coaching and consulting.

Diana's tenure at Pearson stands out in her career. Starting in London as the Vice President of Innovation Partnerships & Head of Future Technologies, she was responsible for trend analysis, strategic partnerships, and early-stage prototype development. She spearheaded initiatives such as an early-stage startup accelerator and cultivated a Future Technologies community within Pearson to drive digital transformation. After returning to San Francisco, she became the VP of Innovative Learning Solutions. Under her leadership, her team generated $86 million in revenue and impacted 11 million learners. Her strategic oversight led to a reduction in delivery time and a significant improvement in the purchase pipeline.

Diana is the instructor of several high-rated courses on product management and leadership, including her cohort-based course on Maven[2] which has graduated four cohorts and has a 4.5/5.0 rating. Her coaching practice emphasizes people-first product leadership, prioritizing the importance of individuals in the product development process. She believes in empowering individuals to reach their full potential, building strong teams, and fostering a culture of innovation and agility.

Diana is a prolific writer, contributor and thought leader on product management, product leadership and leading through a human-centered approach.[3] Interviews and articles showcase her expertise in future technologies and innovation. For example, Wired has featured Diana's information technology and product development insights. She was a finalist for the FDM everywoman in Technology Award in the UK and a featured speaker at product conferences, including Mind the Product, Productized and PDMA Inspire Innovation.

Diana holds a bachelor's degree from the University of California *(UC)*, San Diego. She also earned a Master of Information Management and Systems *(MIMS)* from UC Berkeley and an MBA from Boston University. Her certifications include Business Innovation Foundations, Certified Technical Product Manager, and more. Diana is a member of the International Coaching Federation.

ENDNOTES

Introduction

1. Product School. (2020, September 16). Product management: 90s vs. 2020. Product School. Retrieved June 7, 2024, from https://productschool.com/blog/product-strategy/product-management-90s-vs-2020/

Part I: Product Management is Changing, Better Keep Up

Chapter 1: Product Management Then vs. Product Manageent Now

Chapter 2: The mini-CEO

1. Kevin Systrom (Instagram), Marissa Mayer (Google), Satya Nadella (Microsoft), and Sundar Pichai (Google)

2. Rumelt, Richard. Good Strategy Bad Strategy: The Difference and Why It Matters. New York: Crown Business, 2011

3. Trzeciak, by Stephen; Mazzarelli, Anthony and Seppälä, Emma. "Leading with Compassion Has

Research-Backed Benefits." Harvard Business Review. February 2023. https://hbr.org/2023/02/leading-with-compassion-has-research-backed-benefits

Chapter 3: Everchanging Product Frameworks

1. Pragmatic Institute. *Product Management & Product Marketing Learning Ecosystem.* Pragmatic Institute, https://www.pragmaticinstitute.com/product/#learning-ecosystem. Accessed 25 Sept. 2024

2. Olsen, Dan. *The Lean Product Playbook: How to Innovate with Minimum Viable Products and Rapid Customer Feedback.* Wiley, 2015.

3. Cagan, Marty. *Inspired: How to Create Products Customers Love.* 2nd ed., Wiley, 2018.

4. Torres, Teresa. C*ontinuous Discovery Habits: Discover Products That Create Customer Value.* Product Talk LLC, 2021.

5. Udemy. *Online Courses – Learn Anything, On Your Schedule.* Udemy, https://www.udemy.com/. Accessed 25 Sept. 2024

6. Cagan, Marty, and Chris Jones. *Empowered: Ordinary People, Extraordinary Products.* Wiley, 2020.

8. Maverick, J. B. What Are FAANG Stocks? New Name MAMAA Explained. Forbes, 3 Oct. 2023, https://www.forbes.com/advisor/investing/faang-stocks-mamaa/

Author Reflections: Teresa Cain

1. Lexmark International, Inc. *(n.d.) Lexmark Announces Definitive Agreement to Acquire Perceptive Software.* Lexmark Newsroom. Retrieved on July 1, 2024 from https://newsroom.lexmark.com/newsreleases?item=23833

2. Hyland. *(n.d.) Hyland Finalizes its Acquisition of the Perceptive Business Unit.* Retrieved on July 1, 2024 from https://www.hyland.com/en/company/newsroom/hyland-finalizes-its-acquisition-of-the-perceptive-business-unit

3. Gibbons, S., Moran, K., Pernice, K. & Whitenton, K. Nielsen Normal Group. *(2021, June 13) The 6 Levels of UX Maturity.* Retrieved on December 30, 2023 from https://www.nngroup.com/articles/ux-maturity-model/

Chapter 4: Your Path to Product Management

1. 1. Scrum *(n.d.) The Scrum Guide.* Retrieved on July 1, 2024 from https://www.scrum.org/resources/scrum-guide

Chapter 5: Challenges for Product Managers to Overcome

1. Levels.fyi *(n.d.) Compare Career Levels and Compensation Across Companies.* Retrieved on July 1, 2024 from https://www.levels.fyi/

2. Salary.com *(n.d.) Product Management Manager Salary.* Retrieved on July 1, 2024 from https://www.salary.

com/research/salary/benchmark/product-management-manager-salary

3.   Levels.fyi *(n.d.) LinkedIn Product Manager Salaries.* Retrieved on July 1, 2024 from https://www.levels.fyi/companies/linkedin/salaries/product-manager?country=254

4.   OpenAI *(n.d.) ChatGPT.* Retrieved on July 1, 2024 from https://www.chatgpt.com

Chapter 6: What does it Take to be a Successful Product Manager Today?

1.   Knight, Rebecca. *How to Improve Your Soft Skills as a Remote Worker.* Harvard Business Review. January 8, 2024. Retrieved on July 1, 2024 from https://hbr.org/2024/01/how-to-improve-your-soft-skills-as-a-remote-worker

2.   New York Daily News Staff. *Knowledge Speaks': Jimi Hendrix's Top Quotes on the Anniversary of His Death.* New York Daily News. September 18, 2015. Retrieved on July 1, 2024 from https://www.nydailynews.com/2015/09/18/knowledge-speaks-jimi-hendrixs-top-quotes-on-the-anniversary-of-his-

3.   McGregor, Douglas. *The Human Side of Enterprise, Annotated Edition.* [Kindle edition]. McGraw Hill. Retrieved on July 1, 2024 from https://amazon.com/Human-Side-Enterprise-Annotated-ebook/dp/B009E71XRE

4.   Senge, Peter M. *The Fifth Discipline: The Art & Practice of The Learning Organization.* Kindle edition. New

York: Currency Doubleday, 2006. Retrieved on July 1, 2024 from https://www.amazon.com/Fifth-Discipline-Practice-Learning-Organization-ebook/dp/B000SEI-FKK

5.   Microsoft. *Hybrid Work Is Just Work.* Work Trend Index. Last modified 2024. Retrieved on July 1, 2024 from https://www.microsoft.com/en-us/worklab/work-trend-index/hybrid-work-is-just-work

Part II: Future Proof Your Career as a Next-Gen Product Manager

Chapter 7: Where is the Next Evolution of Product Management?

1.   Fernandez, Jenny and Velasquez, Luis. *Team Building Diversity: Complementary Skills for Success.* Fast Company. June 1, 2023. Retrieved on July 1, 2024 from https://www.fastcompany.com/90823224/team-building-diversity-complementary-skills-success

2.   McKinsey & Company. *Diversity Wins: How Inclusion Matters.* May 2020. PDF file. Retrieved on July 1, 2024 from https://www.mckinsey.com/~/media/mckinsey/featured%20insights/diversity%20and%20inclusion/diversity%20wins%20how%20inclusion%20matters/diversity-wins-how-inclusion-matters-vf.pdf

3.   StartupArchive on X.com. Tweet. Retrieved on July 1, 2024 from https://x.com/StartupArchive_/status/1766083999028441270

Chapter 8: Practice, Don't Preach Product Management Principles

1. "Lady Justice." *Historical Society of the New York Courts.* Retrieved on July 1, 2024 from https://history.nycourts.gov/history-new-york-courthouses/lady-justice/

2. "Medusa." Encyclopaedia Britannica. *Medusa Greek Mythology.* Retrieved on July 1, 2024 from https://www.britannica.com/topic/Medusa-Greek-mythology

3. "Scotty Principle." Urban Dictionary. *Scotty Principle.* Retrieved on July 1, 2024 from https://www.urbandictionary.com/define.php?term=Scotty%20Principle

Author Reflections: Dr. Bart Jaworski

Chapter 9: Be a Product Champion, Not Just an Order Taker

1. Robinson, Bryan. *Discover the Top 5 Reasons Workers Want to Quit Their Jobs.* Stillman. Forbes. 3 May 2022. https://www.forbes.com/sites/bryanrobinson/2022/05/03/discover-the-top-5-reasons-workers-want-to-quit-their-jobs/

2. Stillman, Daniel. *Minimum Viable Transformation.* Daniel Stillman. Retrieved on July 1, 2024 from https://www.danielstillman.com/blog/minimum-viable-transformation

3. Scentola, Damon. *Tipping Point for Large-Scale Social Change.* Penn Today. Retrieved on July 1, 2024 from https://penntoday.upenn.edu/news/damon-centola-

tipping-point-large-scale-social-change

Chapter 10: Drive More Results with Data

Chapter 11: Using AI Tools to Stay Sharp

1. *Is Nvidia Corporation (NVDA) A Hedge Fund Favorite?* 23 Sept. 2023, Retrieved on July 1, 2024 from https:// www.finance.yahoo.com/news/nvidia-corporation-nvda-hedge-funds-123554634.html

2. *How Many AI Companies Will There Be in 2024? Latest AI Statistics.* Springsapps, Retrieved on July 1, 2024 from https://www.springsapps.com/knowledge/how-many-ai-companies-will-there-be-in-2024-latest-ai-statistics

3. OpenAI *(n.d.) ChatGPT.* Retrieved on July 1, 2024 from https://www.chatgpt.com

4. Grammarly *(n.d.) Grammarly.* Retrieved on July 1, 2024 from https://www.grammarly.com

5. Orai *(n.d.) Orai.* Retrieved on July 1, 2024 from https://www.amplitude.com

6. Amplitude *(n.d.) Amplitude.* Retrieved on July 1, 2024 from https://www.orai.com

7. FigJam *(n.d.) FigJam.* Retrieved on July 1, 2024 from https://www.figma.com/figjam

Part III: Product Manager to Product Leader

Chapter 12: Success for a Product Manager vs. Product Leader

1. Maxwell, John C. T*eamwork Makes the Dream Work.* Advantage Quest Publications, 2008.

Chapter 13: You Don't Need to be Loud to Lead

1. Abramson, Ashley. *How to Overcome Impostor Phenomenon.* APA Monitor on Psychology. June 2021. https://www.apa.org/monitor/2021/06/cover-impostor-phenomenon.

2. Duckworth, Angela. Grit: The Power of Passion and Perseverance. New York: Scribner, 2016.

3. *The Link Between Personality and Success.* The Economist. March 18, 2021. https://www.economist.com/business/2021/03/18/the-link-between-personality-and-success

Author Reflections: Diana Stepner

1. McKinsey & Company. *Enduring Ideas: The Three Horizons of Growth.* Strategy & Corporate Finance Insights. Retrieved on July 1, 2024 from https://www.mckinsey.com/capabilities/strategy-and-corporate-finance/our-insights/enduring-ideas-the-three-horizons-of-growth

Chapter 14: Using Your Leadership Style to Deliver Results

1. *Why Am I Talking? The Power of W.A.I.T.* Center for the Empowerment Dynamic. Retrieved on July 1,

2024 from https://theempowermentdynamic.com/why-am-i-talking-the-power-of-w-a-i-t

2.   Wilson, Fred. *Be Generous.* AVC. Retrieved on July 1, 2024 from https://avc.xyz/be-generous

Chapter 15: Managing Stakeholders at every Level of the Organization

1.   Arekmalang. A Stack of Raw Steak on a Table with a Fork and a Knife. Adobe Stock. https://stock.adobe.com/photo/a-stack-of-raw-steak-on-a-table-with-a-fork-and-a-knife-123456789. Accessed 17 Nov. 2024.

2.   Gooding, Grant. *Henry Ford Innovation Paradox - Solved.* Thinking Bigger, March 9, 2021. Retrievedon July1, 2024 from https://ithinkbigger.com/the-henry-ford-innovation-paradox-solved/

Conclusion

1.   *Young Leaders.* CoGenerate. Retrieved on July 1, 2024 from https://cogenerate.org/young-leaders/

2.   Twenge, Jean. Generations: The Real Differences Between Gen Z, Millennials, Gen X, Boomers, and Silents—and What They Mean for America's Future. Atria Books, 2023.

About Teresa Cain

1.   Lucid Startup Consulting. *Home.* https://www.lucid-startupconsulting.com

2. Cain, Teresa. *Home.* Teresa Cain. https://www.teresa-cain.com

3. Udemy. *Instructor.* Teresa Cain. https://www.udemy.com/user/teresa-cain-3/

4. Cain, Teresa. *Solving Problems in 2 Hours: How to Brainstorm and Create Solutions with Two-Hour Design Sprints.* 2nd ed., Lucid Creative Press, 2024, https://a.co/d/b6MtOMJ

About Dr. Bart Jaworksi

1. Jaworski, Bartosz. *Home.* Medium. https://dr-bartosz-jaworski.medium.com/

2. Jaworski, Bartosz. *Bartosz Jaworski, Author Page.* LogRocket Blog, https://blog.logrocket.com/author/bartoszjaworski/

3. Jaworski, Bartosz. *Home.* Bartosz Jaworski. https://www.drbartpm.com

About Diana Stepner

1. Stepner, Diana. *Home.* Diana Stepner. https://dianas-tepner.com

2. Stepner, Diana. *Home.* Diana Stepner. https://maven.com/diana-stepner/product-leadership

3. Stepner, Diana. *Home.* Substack. https://dianastepner.substack.com/

SPECIAL THANKS

Special thanks to all of our colleagues, friends, family, early-readers, followers, entrepreneurs and organizations in the product management community for your support and for sharing the same passion and vision as us for next-gen product management.

Teresa, Bart, Diana